Making Sense of Expertise

Current debates about experts are often polarized and based on mistaken assumptions, with expertise either defended or denigrated. *Making Sense of Expertise* instead proposes a conceptual framework for the study of expertise in order to facilitate a more nuanced understanding of the role of expertise in contemporary society.

Too often, different meanings of experts and expertise are implied without making them explicit. Grundmann's approach to expertise is based on a synthesis of approaches that exist in various fields of knowledge. The book aims at dispelling much of the confusion by offering a comprehensive and rigorous framework for the study of expertise. A series of in-depth case studies drawn from contemporary issues, including climate change, the Covid-19 pandemic, and the challenge to professional expertise provide the empirical basis of the author's comprehensive approach.

This thought-provoking book will be of great interest to students, instructors, and researchers in a range of fields in the humanities, social sciences, and science and technology studies.

Reiner Grundmann is a Professor of Science and Technology Studies at the University of Nottingham, UK. His main research interests lie in the relationship between knowledge and decision-making. In recent years, he has studied the public discourse on climate change and how scientific experts, decision makers, and civil society groups have framed the issue in the mass media. Grundmann also researches issues in social theory and has published on Karl Marx, Werner Sombart and Otto Neurath. He is the author of *Transnational Environmental Policy: Reconstructing Ozone* (Routledge, 2001) and co-author of *Experts: The Knowledge and Power of Expertise* (Routledge, 2011) and *The Power of Scientific Knowledge: From Research to Public Policy* (Cambridge University Press, 2012). He is also chief editor for the specialty section Climate and Decision Making in the *Frontiers in Climate* journal and co-editor of *Economic Life in the Modern Age* (Transaction, 2001), *Knowledge* (five volumes, Routledge, 2005), and *Society* (four volumes, Routledge, 2008).

'This book is a sophisticated attempt to provide a theory of expertise that defines it as a category in the social and organizational worlds. It engages a variety of literatures that are segregated from one another, which students should be exposed to. The discussions of different traditions in the expertise literature, in different chapters, serves any reader well.'

Stephen Turner, author of The Politics of Expertise *(Routledge, 2013), and Distinguished University Professor of Philosophy and Director of Center for Social & Political Thought, University of South Florida, USA*

'In this book Grundmann offers new theoretical insights into the nature and operation of expertise in society. He combines his theoretical framework with an empirical application to highly relevant examples, making fascinating discoveries along the way. He is analytically sharp and applies his insights to cases that are highly relevant in current debates, such as Covid-19, Climate Change, or Artificial Intelligence. He manages to resolve the confusion surrounding the role of expertise in these debates. The reader is rewarded with new insights, based on a ground-breaking approach.'

Nico Stehr, author of Knowledge Capitalism *(Routledge, 2022) and Emeritus Karl Mannheim Professor of Cultural Studies, Zeppelin University, Germany*

Making Sense of Expertise

Cases from Law, Medicine, Journalism, Covid-19, and Climate Change

Reiner Grundmann

Routledge
Taylor & Francis Group
NEW YORK AND LONDON

Cover image: Rebecca Grundmann, 2004

First published 2023
by Routledge
605 Third Avenue, New York, NY 10158

and by Routledge
4 Park Square, Milton Park, Abingdon, Oxon, OX14 4RN

Routledge is an imprint of the Taylor & Francis Group, an informa business

Library of Congress Cataloging-in-Publication Data
Names: Grundmann, Reiner, author.
Title: Making sense of expertise / Reiner Grundmann.
Description: New York, NY : Routledge, 2023. | Includes bibliographical references and index. | Identifiers: LCCN 2022022754 (print) | LCCN 2022022755 (ebook) | ISBN 9781032335674 (hardback) | ISBN 9781032335643 (paperback) | ISBN 9781003320227 (ebook)
Subjects: LCSH: Expertise. | Specialists.
Classification: LCC BF378.E94 G86 2023 (print) | LCC BF378.E94 (ebook) | DDC 153.9--dc23/eng/20220729
LC record available at https://lccn.loc.gov/2022022754
LC ebook record available at https://lccn.loc.gov/202202275

ISBN: 978-1-032-33567-4 (hbk)
ISBN: 978-1-032-33564-3 (pbk)
ISBN: 978-1-003-32022-7 (ebk)

DOI: 10.4324/9781003320227

Typeset in Garamond
by KnowledgeWorks Global Ltd.

To the memory of Emma Grundmann (1926–2017)

Contents

List of Table

List of Acronyms

AI	Artificial Intelligence
ANT	Actor-Network-Theory
AR	Assessment report
BECCS	Bioenergy with carbon capture and storage
COP	Conference of the Parties
COVAX	COVID-19 Vaccines Global Access
GHG	Greenhouse gas
IPCC	Intergovernmental Panel on Climate Change
NHS	National Health Service
NPI	Non-pharmaceutical intervention
PNS	Post-normal Science
RCP	Representative Concentration Pathway
SAGE	Scientific Advisory Group for Emergencies
SBSTA	Subsidiary Body for Scientific and Technological Advice
SPM	Summary for Policymakers
STS	Science and Technology Studies
UNEP	United Nations Environment Programme
UNFCCC	United Nations Framework Convention on Climate Change
WHO	World Health Organization
WMO	World Meteorological Organization

Acknowledgements

Parts of this book incorporate material which was previously published. I am grateful for the publishing houses for granting the right to reprint these sources in part or completely.

Chapter 2: Grundmann, Reiner. 2017. 'The Problem of Expertise in Knowledge Societies.' *Minerva* 55(1): 25–48..

Chapter 4: Grundmann, Reiner. 2018.'Knowledge Politics.' In *The Blackwell Encyclopedia of Sociology*, Oxford, UK: John Wiley & Sons, Ltd.

Grundmann, Reiner, and Nico Stehr. 2012. *The Power of Scientific Knowledge: From Research to Public Policy*. Cambridge: Cambridge University Press.

Chapter 9: Grundmann, Reiner. 2021. 'COVID19: Expertise, Transparency and the Problem of Trust.' *Discover Society* 1(3).

Grundmann, Reiner. 2021. 'COVID and Climate: Similarities and Differences.' *WIREs Climate Change* 12(6).

Chapter 11

Grundmann, Reiner. 2018. 'The Rightful Place of Expertise.' *Social Epistemology* 32(6): 372–86.

I have greatly benefited from the comments by Steve Fuller, Dimitris Papadopoulos, Nico Stehr and Stephen Turner. Their astute observations helped improve the manuscript considerably. Oliver Geden and Hans von Storch have seen a previous version of chapter 7, The IPCC: A Chameleon of Expertise and provided me with some really valuable comments. All remaining shortcomings are my own.

Introduction

There is a lot of talk about the crisis of expertise, some authors even pronouncing the death of expertise. The problem with such statements is that they are not specific enough. They make sense with regard to particular forms of expertise but are not valid as a general diagnosis of our time. To give the reader a flavour of what is to come, I will only say this much at the outset: what can be described as 'official expertise' has been challenged. At the same time, more sources of expertise have emerged. Both trends are linked, and the former should be seen as a consequence of the latter.

The book has three parts. Part I puts forward a conceptual framework for the analysis of expertise and experts in contemporary societies. This is based on a review of relevant contributions from Science and Technology Studies (STS), the sociology of the professions, political psychology, economics, and political science. Part II examines several empirical cases in the light of the conceptual framework. Part III discusses the consequences and implications for society.

In *part I,* I show that prevailing approaches have come to see expertise either as linked to a profession or to science. These approaches tend to neglect the wider historical and social context in which expertise operates, and the differences between political cultures across national jurisdictions. I will address these shortcomings and develop a framework drawing on approaches from several social science contributions.

The framework of expertise presented here is intended to show the limitations of existing frameworks and the gain in knowledge we will get by viewing expertise in its social and historical context. A new typology will be introduced, which distinguishes between different roles of experts in society.

This book builds on two co-authored books with Nico Stehr (Grundmann and Stehr 2012; Stehr and Grundmann 2011), and two journal articles (Grundmann 2017, 2018). In these publications, I examined the link between knowledge and decision-making and suggested to critically examine existing concepts of expertise. This work has led to a definition of expertise as relational, situated in a social context where a client requests expertise from an expert who is trusted to have the requisite knowledge and experience.

DOI: 10.4324/9781003320227-1

I have also suggested to include the role of the *specialist* who is someone who has the technical knowledge and is empowered to apply it without mediation or a special relationship with a client (Grundmann 2018). Specialists are a much wider group of people than professionals (but cp. Abbott 1988). Specialists need not be members of professional organizations although many are.

Here I will go beyond these insights and extend the typology. First, we need to add the role of the *commentator* who offers opinions based on a close familiarity with a subject matter. In contrast to the specialist, who has quasi-jurisdictional powers, commentators have no direct power. They have indirect power through framing an issue and influencing a discourse. Their role is to assess, select and evaluate a range of information for the general public. They do not stand in a direct relation to a client; it does not make sense to conceptualize the general public as the client as it cannot articulate specific requests for guidance, although such requests are common. Commentators provide various kinds of information to their audiences, ranging from highly specialized technical knowledge to very general orientation knowledge, as provided, for example, by public intellectuals.

A theory of expertise would be incomplete if it did not include non-human forms of expertise, especially *artificial intelligence* (AI). These 'expert systems' provide impersonal yet customized advice and suggest or even enact a course of action. These systems are embedded in everyday life, from weather forecasts to traffic reports. They have been proliferating and could become ubiquitous. They provide guidance for action for individuals and organizations, and in some instances, replace the service provided by human experts. These systems are part and parcel of the availability of 'big data' and its affordances (Mayer-Schönberger and Cukier 2013).

Some AI systems are also able to replace human action directly, as is the case with robots. Robots are operating according to a logic which follows the deskilling of human labour whereby a mechanism assumes the role of human activity. Robots additionally incorporate software which makes them function the way they do. In analogy to the distinction between advisors and specialists among humans, we can add the same distinction for AI systems: there are systems that provide advice to human decision-makers, and there are devices that make and implement a decision themselves. Traffic guiding systems and cars equipped with electronic systems are an instance of the former, self-driving cars would be an instance of the latter.

The social science literature on AI is expanding fast. So far, legal and medical applications have attracted the most attention. For example, in the field of law and policing, AI systems have been used to assess the likelihood of recidivism (Cobbe 2019; Oswald 2018), in medicine, the screening of X-rays to detect pathologies is prominent (Ho et al. 2019). Without much exaggeration, one can say that crime and cancer have been major reasons for developing these applications. In the field of medicine, such developments have been welcomed based on their alleged higher accuracy compared to humans (Grove and Meehl

1996; see already Meehl 1954). Others have pointed at deeply problematic features, which include 'discrimination by code,' wrong predictions and socially unwanted outcomes (Angwin et al. 2016; see Forrest 2021 for a discussion).

Theoretically, AI systems pose the question of agency. Who is responsible if something goes wrong? The machine, the company that runs it, the user, the programmer? Actor-network theory has a general answer about distributed agency but is not fine grained enough to deal with the questions raised (Latour 2005). Sociologists of technology and law have suggested to use a differentiated, gradual concept of agency and of legal responsibility (Rammert 2008; Teubner 2006).

I will draw on these contributions to develop a typology which provides a much-needed conceptual clarification. It will go beyond the existing literature, which tends to limit expertise to either scientists (e.g. Roger Pielke's Honest Brokers), to science advisory bodies (STS more generally), or to professionals (e.g. Andrew Abbott). The proliferation of citizen or 'field' expertise and of AI systems widens the notion of expertise, as does the recognition of the specialist and the commentator. The thesis of the book is that we have seen a rise of the role of the commentator and an expansion of AI systems. In some instances, both combine to form a pincer movement against established forms of expertise.

In *part II,* I discuss empirical cases using examples which have characteristics that call for an application of the typology. The usefulness of the typology (and its limitations) will be discussed. It is expected that some counter-intuitive results follow from this discussion. I will examine two salient and topical global problems and three fields of professional expertise. These cases have different properties: the important aspects of Climate change and Covid-19 are policy related, whereas professional expertise is interesting because it is the prime example where the pincer movement is evident.

I will start with the example of the Intergovernmental Panel on Climate Change (IPCC) and ask what kind of expertise it provides and what role experts play. I explore the thesis that the IPCC is an intergovernmental body which is commonly said to provide expert advice to governments but, in fact, does not do so. It is prevented from playing this role through its own remit, which is to assemble and evaluate scientific research, and to stay policy neutral. Climatology, the study of physical properties of climate, is the leading discipline. It provides the epistemic core for framing and understanding the issue. It has built a global knowledge infrastructure, including climate models. It has produced systems of classification and unified metrics. Perhaps most importantly, the IPCC and commentators speaking in its name, or criticising it, provide comments to a world audience, framing the problem in specific ways. But these are often emphasizing different aspects or contradicting each other which leads to a contest about epistemic authority.

Similar to climate change, the Covid-19 pandemic has put the role of expertise into sharp relief. Political decisions, involving far-reaching restrictions to social life, have been justified on the basis of epidemiological models. The practical impact of this expertise has been fast and sweeping. While some systems of classifications have been developed, the lack of unified data and metrics is noteworthy. Nevertheless, policies are justified on the basis of data. Commentators who go against the official epidemiological expertise are often belittled or stigmatized as conspiracy theorists.

I will then examine expertise based on the professions, looking at cases that have been prominent in previous research: medicine, law, and journalism. While medicine and journalism have come under double pressure from educated citizens and technological developments, law seems to be less affected by this pincer movement.

Part III will discuss the implications of the results from parts one and two. I argue that the multiplication of knowledge sources has led to an erosion of official knowledge providers and experts and also to an erosion of trust in mainstream institutions of society. But it also has led to a process of democratization in which knowledge (and truth claims) are not so much received through trusted authorities but via a complex ecosystem of competing voices, making claims and offering their services.

I will discuss the contours of the knowledge society which take shape as a result of these developments. What kind of social change is on the cards? Is technology driving social change via AI applications, or is society adopting selected applications in a limited manner? Are populist leaders abandoning the institutional and professional environment which provides expert advice for governing? Will lay publics become more and more expert? Is the 'death of expertise' a real threat? What does this mean for democracy?

Throughout the book, I will operate a musical metaphor, referring to two different types of music. The first is based on a simple riff and a long coda, think of the Beatles' song *Hey Jude*. The riff goes like this: in most expert debates, there are real experts and fake experts. The real experts need to be trusted because they are neutral academic or government experts. They have a consensual view based on scientific practices. The fake experts are not to be believed, because they are operating outside the consensual view. This riff often has a long coda, dwelling on the outsiders' illegitimate funding or their ideological motives. Those who play this song or like to listen to it see the problem of epistemic trust as a problem of good and bad, even of good and evil. Maybe the most prominent representative of this tune in the social sciences is Naomi Oreskes. In her book *Why Trust Science,* she writes:

> But it is in the nature of expertise that we trust experts to do jobs for which they are trained and we are not. Without this trust in experts, society would come to a standstill. Scientists are our designated experts for studying the world. Therefore, to the extent that we should trust anyone to tell us about the world, we should trust scientists.
>
> (Oreskes 2019:56)

This puts scientists in a monopoly position to produce knowledge about the world, which is a questionable proposition. To be fair to Oreskes, she modifies her position when discussing non-experts' potential for contributing knowledge. She reviews some contributions from STS and agrees with Brian Wynne that laypersons are not epistemically vacuous. However, she then goes on to qualify the role of non-experts, especially in situations where they start criticizing 'the science':

> When people without relevant expertise criticize science, we should consider the possibility that something fishy may be going on. If people are attacking science, there is something at stake, but it is not necessarily something scientific. Indeed, it is probably not.
>
> (Oreskes 2019:132)

Attacks on 'the science' are explained by the nefarious activities of organized business interests, main examples are the tobacco and fossil fuel industries. Science has reached a consensus view about the risks associated with smoking tobacco and burning fossil fuels, but misinformation campaigns by interested parties want to make it look as if there was no consensus. This is a rosy picture of science and a simplistic account of science controversies which have been discussed elsewhere in greater detail (Pearce et al. 2017; Grundmann 2013, 2015).

What matters in this context, and what Oreskes does not mention, is the fact that when we are dealing with ill-defined problems and need to reach a decision, science is of little help. I am not saying it is of no help, but to think that 'we should trust scientists' because they are the experts to tell us about the world is a deeply flawed idea. Not only scientists are observing the world, everyone else is too. What is more, under conditions of uncertainty, their contribution is one among many, and they sometimes turn out to be wrong.

In such debates, we hear another tune, a rather dissonant and provocative one, think of Jimi Hendrix's *Star-Spangled Banner*. It amounts to a deconstruction of an official rendition, but does not provide a simple riff or a neat tune. We rather get a jarring set of intonations that highlight the many different aspects of expertise, including a brief recognition of the official version. If there is one common sound that unites these different layers, it is the scepticism vis-à-vis claims about the trustworthiness of official expertise. Scholars in various fields in the social sciences, including STS, economics, psychology, and political science, are intoning variations on this theme. However, some early proponents of this tune have changed their repertoire and developed a preference for the simple tune, even if the coda in their version of *Hey Jude* is subdued.

Part I

Chapter 1

A General Concept of Expertise

Useful definitions matter. They matter not only because they help us understand the world around us but to guide empirical research giving us new insights, they matter also as an antidote against an unspecified and often moralizing discourse on experts, which in the past years has become either a blanket defence of expertise or a wholesale attack. In this sense, they help us to approach practical-political problems from a different vantage point.

Knowledge-based expertise

However, before embarking on the concept of expertise, I want to clarify the meaning of the term knowledge. Since expertise is based on knowledge, skills, and experience, a working definition of knowledge is required. There are forms of expertise where skills and experience are much more relevant, as is the case with sports persons or artists. All forms of expertise comprise the three elements of knowledge, skill, and experience, but the relative importance varies.

In a general sense, I define knowledge as information about the world. This is the cognitive aspect of knowledge. Some of this information can be used to make a difference in the world. This is the practical or instrumental aspect of knowledge. The cognitive dimension of knowledge contains factual, credible, or certified information, that is, knowledge we rely upon. It also contains uncertain knowledge, which either will be transformed into certain knowledge or remain uncertain. If we want to or have to make a decision, this situation usually calls for expertise.

The practical aspect of knowledge can be based on publicly available knowledge or on esoteric knowledge. The latter often confers an advantage vis-à-vis nature and vis-à-vis others. This advantage is based on speed and secrecy. Competitive advantage is often knowledge based, and it rests on the application of some information that others do not have at any given point in time. The growing importance of experts can be seen in this especially interesting nexus between uncertain knowledge and action oriented counsel for decision-making. Experts are especially relevant if they can provide advice, and legitimate decisions in situations where knowledge is uncertain and multiple

DOI: 10.4324/9781003320227-3

options for action exist. Nico Stehr famously called this property of knowledge 'the capacity to act.' It was meant to emphasize that knowledge as such is not power but provides the conditions and resources to intervene in reality and to make a difference.

People, including social scientists, often use the terms expert and expertise in a general way without paying attention to definitional questions. Experts are seen as scientists or professionals, and sometimes the label is given as a badge of honour to flatter an interlocutor. But experts have also been criticized for being self-serving and abusing their privileged position in society. I am convinced that at least some of these differences are due to the fact that different aspects of expertise underly the different evaluations, leading to confusion.

The argument developed here builds on my own work, partly developed with Nico Stehr. We have emphasized a relational definition of expertise that can be summarized as follows: experts mediate between the production of knowledge and its application; they define situations; they deal with value conflicts; and they set priorities for action. Experts are judged by clients, not by professional or scientific peers; and they rely on trust by their clients. Experts combine facts with opinion, knowledge about facts and uncertainties with interpretation, and most crucially with an action orientation. They advise clients on how to act in the absence of certain knowledge. This relational definition holds that we should see experts as consultants who advise decision-makers (individuals and organizations), making recommendations that can be put into practice. This is my starting point which is based on previous work (Grundmann 2017; Grundmann and Stehr 2012; Stehr and Grundmann 2011).

Advisors, specialists, commentators, and artificial intelligence

Defining experts as advisors captures important aspects of expertise but needs to be expanded. We need to include those types of experts who do not stand in a direct relation to a client (Grundmann 2018). These may be serving a principal or offering their opinion to a wider audience, not a specific client. They may be human or non-human (more-than-human). Three roles are important here: the specialist, the commentator, and expert systems (aka artificial intelligence, AI).

The specialist, who often is a professional, is trained to deal with complicated situations and has a lot of knowledge and competence acquired through training and experience. Often a specialist has the power to make decisions as their professional role allows and demands. Examples are engineers, lawyers, doctors, pilots, or teachers. But also, members of private and public organizations who do not have professional accreditations fall into this category. The specialists' activity can be understood as the skilful application of rules to problems. Specialists command a large body of technical knowledge and are able to apply it in practical contexts. Normally they implement a course of action as part of their daily duties and because their authority allows and demands them to do so. They tend to be overseen by a principal.

The commentators are experts who do not have a client to report to and are not overseen by a principal. Commentators do not advise clients and do not implement decisions—they are providing their opinions and reflections to a wider audience. Examples are sports commentators, political commentators, music critics, public intellectuals, and, not to forget, also scientific commentators.

Some technological systems are set up to compute assessments and forecasts, which can be used as a guide for action. They are products of human knowledge, skills, and experience but have taken on a role of their own, exercising some kind of agency. Like the advisor, expert systems can be seen as standing in relation to a client. And automated expert systems such as robots can implement decisions without further mediation. It is still an open question if they will be able to provide comments to a wide audience.

Only advisors and commentators are dealing with uncertainty and value conflicts. It is expected that they will weigh the different possible merits and potential outcomes of different courses of actions and provide their conclusion. Advisors do this for their clients who pay them, commentators do this to the wider public. Specialists are not in the business of addressing uncertainty or value conflicts. AI systems such as robots could be 'programmed' to deal with value conflicts and uncertainty, even if the process of decision-making is hidden and not transparent. The opacity of these systems poses a problem as they, unlike human experts, cannot be challenged and interrogated.

Table 1.1 presents a typology of experts. I am distinguishing between cognitive and practical tasks and between well-defined and ill-defined problems. These terms are alluding to the distinction between tame and wicked problems, which are practical-political in nature (see Peters 2017; Simon 1973). There are also well-defined cognitive problems which scientists deal with. The box of ill-defined cognitive problems is empty, as scientists always strive to define research problems in such a way that they can be solved. One could imagine several types of actors dealing with ill-defined problems, with many scientists trying to abstain from what they typically perceive as a foolish endeavour. One needs, a bit like Maxwell's demon, constantly try to transform ill-defined into well-defined problems. Where scientists do engage with such problems they aim to shift them 'upstairs' into the NE quadrant while at the same time avoiding stuff from upstairs seeping down.

Table 1.1 Typology of Experts

	Cognitive	*Practical*
Well-defined problems	Scientist	Specialist, AI
Ill-defined problems	?	Advisor Commentator

The roles of advisor and commentator can be played by issue advocates and honest brokers (to use Roger Pielke's terminology). And both roles can be played by citizen experts, not only by scientists or by professional experts. The main difference between advisors and commentators is that the former work for a client while the latter communicate in public. The former tend to be close to the client and often work behind closed doors. The latter participate in public debates. Both give opinions, but advisors make recommendations for their clients, too.

Interesting cases appear when government advisors or scientists assume the role of commentator. This is a tempting move, and often they get away with it because they are perceived according to their main role (scientist, government advisor) and thus exploit a reputational surplus which can help them gaining public credibility or epistemic authority.

Last but not least, there is an important difference in the way the cognitive and practical relate to each other. In well-defined problems, the impression is that there is a direct translation from the world of knowledge to the world of action. The linear model of science policy relations has its rational kernel here. But it is a fallacy to generalize this situation to a situation where we are dealing with ill-defined problems. A linear relation in this situation would only be possible if scientists could somehow master ill-defined problems.

Oblivious of this fact, scientists make claims about managing uncertainty, wanting to appear as being in a position to make well-founded recommendations. However, their claims are an appeal to their social status, which supposedly confers epistemic authority.[1] In reality, they are still dealing with ill-defined problems, an inconvenience which tends to be glossed over. In this story, we are presented with a linear relation between facts and decisions as the general model for science policy relations.[2]

This logical fallacy is part of a rationalist ideology which cannot be reduced to any kind of bad intention. The temptation is strong to use this scheme of interpretation whenever scientists find themselves confronting ill-defined problems. As my case studies show, the attempts at boundary-work (Gieryn 1995) are not always successful.

Public curiosity

Various scholars have commented on human specialists. Max Weber was one of the first social theorists thematizing technical artefacts in an essay from 1913 on interpretive Sociology. Somewhat uncharacteristically for him, he writes about the features of an electric tram, a hydraulic lift and a gun and states that the normal user does not know anything 'about the scientific principles underlying their construction.' He then generalizes this insight to our everyday use of artefacts:

> No ordinary consumer will nowadays have even a rough knowledge of the production techniques of the goods that he uses daily, and he will mostly

> not even know what materials they are made of and what industry has produced them. All that he is interested in are those expectations concerning the performance of these artefacts that are of practical importance for him. The situation is no different with respect to social phenomena – money, for example.
>
> (Weber 2012:300).

Mark Warren, in a similar vein says: 'we want safe airplanes and food, not the chance to participate in meat inspection and airline safety. We wish, perhaps, to participate to the extent that we are assured the rules and routines regulating such matters will put to best use the expertise of those responsible. But most of us are quite willing to trust authorities for the daily decisions that make these sorts of things work. Limited time, large scale, complexity, and interdependency mean that–as generic social relationships–decisions taken on authority flowing from trust are more important now than ever before' (Warren 1996:49).

Our willingness to trust means that we move our doubts into the background, where they become invisible and silent (but do not disappear altogether). Thus Bruno Latour observes: 'The silent working of the sewage systems in Paris has stopped being political, as have vaccinations against smallpox or tuberculosis. It is now in the hands of vast and silent bureaucracies that rarely make the headlines' (Latour 2007:817).

However, as soon as problems appear, value conflicts and uncertainties come to the fore again, and the calm and orderly acceptance of technical artifacts and social institutions becomes challenged. No longer is it the domain of the specialists but a question for politics, and hence for potentially everyone:

> As to the distribution of gender roles, it has been so thoroughly 'naturalized' that it seems at first to be totally outside politics. Should we abstain from calling those issues political in another sense of the adjective? Of course not, because not only did they used [sic] to be loudly disputed controversies (…), but also because they might reopen at any moment, as is clear with the two examples of sewage systems and vaccination above.
>
> (Latour 2007:817).

It seems as if Weber could not envisage this idea in 1913, but a century later, after the expansion of technical systems and new products which have brought comfort, wealth, but also risks, and above all, after the rise of the knowledge society in the last half-century, we do witness at times intense public interest in these things. A concept of expertise needs to be able to theorize these different aspects and dynamics: when are things and institutions taken for granted, and when do they become the object of inquiry and contestation? My preliminary answer is: When they are taken for granted, the specialists reign; when they become contested, advisors are called for, and commentators offer their opinions.

However, the answer gets a bit more complicated as soon as we take citizens into account who may demand a place at the table and as soon as we realize that technology may have already taken a seat. We also need to consider the supply and demand of expertise, something Stephen Turner (2001) has written about.

Supply and demand of expertise

Looking at the supply side, Turner gives examples from economists, theologians, public administrators, and social workers, ranging from single intellectuals to think tanks. On the demand side, he identifies governments ('high politics'), public administration ('bureaucratic apparatus'), NGOs, business firms, and citizens. This leads to a typology of expertise based on different modes of claiming and requesting cognitive authority. Perhaps the most important issue here is the expert's relation to an audience and clientele. It is crucial who pays for the expertise and how clients come to trust it. It is also appropriate to distinguish between experts that advise 'high politics' and those that provide advice to other social users, such as charities or general audiences.

Nevertheless, Turner (2003) also thinks that we cannot conceive of laypeople being on the same level as experts: experts know more than laypeople, and this should be accepted as a fact. While this asymmetry persists logically and ontologically, Turner claims, we can and should design procedures in which expertise is challenged from various corners:

> We are "experts" about our own experience and perhaps about a number of other facts about the world that experts are ignorant of and can usefully be made aware of. But the asymmetry is still there. ... common sense is not inaccessible to experts, as expertise is to common sense.
>
> (Turner 2003:64).

There can be no doubt that this asymmetry exists, it is a logical fact and a product of the definition. While Turner grants that the expert's role is different to the scientist's, he still models the expert closely to esoteric scientific knowledge. At least in this account, experts are closer to scientists than to consultants:

> Just as with the lawyer or the physician, we are dependent on experts because they have a full understanding of the paradigm cases from which they reason analogically. "Dialogue" with them does not make us in any useful sense their equals or discursive partners. We are not even in a good position as non-experts to disagree with the casuistic steps in their reasoning. The construction of analogies is dependent entirely on judgments about what is relevant and what is not that have been acquired by operating from an expert understanding of the paradigm cases. Experts learn

> what differences are relevant to extending causal reasoning on the basis of, and by, possession of the fullest understanding of the cases. As consumers we may learn that experts have their limitations when they perform these feats, as Wynne points out. And we may learn this very painfully. But this makes us better consumers of expertise rather than equal or symmetric discursive partners of experts.
>
> (Turner 2003:63–4).

Turner uses the term consumer thereby somewhat limiting the possible useful contribution of the layperson in terms of providing expertise. This limitation does not seem to follow logically from his premise but from the examples chosen. We are all dependent on experts. Sometimes, people move into the position of expert, while at other times, they are a layperson. Turner does not expand on this, probably because he assumes experts have some kind of professional status. This would limit the definition of expert to professionals, a group that can self-certify. It does not extend to experts in the sense of consultants who are defined and certified by their clients.

The example of medical doctors is instructive. They are part of the medical profession. However, it is worth noting that they do not necessarily produce the knowledge they are using when treating patients. They are not scientists but medical experts (in the sense of consultants—in English, a specialist doctor is called appropriately *consultant*). At this point in the argument, it should become clear that the role of the advisor is not only different to the role of the scientist but also distinct from the role of a professional. These roles may overlap, and scientists may, in fact, be playing all roles at the same time—but this cannot be assumed to be the rule.[3]

Using a relational definition of expertise as mediating between knowledge production and knowledge application, a conceptual space emerges for non-scientists and non-professionals as expert advisors. But this emerging expertise is not necessarily part of knowledge production; it is mediating between knowledge creation and application. Turner's dichotomy between producers and consumers of knowledge seems to obscure this third category, which we could call the excluded middle.

In addition, there are experts with appropriate training who apply their knowledge directly, without being in relation to consumers of knowledge. In other words, what matters in terms of expertise is the fact of making decisions, not only of consuming knowledge.

To conclude, it seems that a useful concept of expertise should examine types of expertise on the supply and demand side. The difference between trained and experienced experts and non-experts is evident but should not make us forget the fact that laypeople can influence decision-making in decisive ways and that specialists can implement decisions directly.

The argument so far lays out that the emergence of the knowledge society has led to a proliferation of and dependence on expertise. On the supply side, there are different forms of expertise on offer, including non-human forms of expertise. On the demand side, more expertise is requested, and the range of requesting actors includes politics, private companies, NGOs, and citizens (Brown 2009; Turner 2001).

Experts as advisors offer guidance and make recommendations to individuals and to organizations. Experts as specialists are applying rules to cases, implementing their own decisions. They are working for a principal. Experts as commentators are offering opinions to a wider audience. Their social position is less circumscribed compared to the advisor and specialist. Commentators come from different fields and have a variety of experience and training. Their relation to the public is what matters. Opinions are sought and provided from a vast range of voices. It is the rise of the commentator role which has led to a structural change in public debates about controversial issues in society.

Experts and narratives

Experts in their role as commentator often appear in public debates where they take a stance. These debates tend to be fraught with moral and political overtones and official experts typically wish to distinguish their contribution as based on knowledge, science, or evidence. However, numbers do not speak for themselves. They need to be embedded in a story that gives meaning to them so that the claims which are made can be credibly supported (Franzosi 1998). Putting numbers into stories is a task that experts often face and often approach in an intuitive way. Metaphors are picked and deployed, and narratives are developed with a timeline, often full with drama, with heroes and villains, even with good and bad substances (Shapin 2007). As Hayden White remarked, 'every historical narrative has as its latent or manifest purpose the desire to moralize the events of which it treats' (White 1980).

Narrative analysis has been developed in policy analysis where four distinctive features have been identified: (1) a setting or context which outlines the problem; (2) a plot with beginning, middle, and end; (3) characters such as heroes, villains, or victims; and (4) a policy solution (Jones and McBeth 2010; Stone 2002). In a similar way Stone suggests that 'Definitions of policy problems usually have narrative structure; that is, they are stories with a beginning, middle, and an end, involving some change or transformation. They have heroes and villains and innocent victims, and they put forces of evil against forces of good' (Stone 2002:130). We can summarize this narrative model in terms of framing, plot, characters, drama, and morals.

The causal and the moral elements of narratives often overlap, are even used synonymously. Think of a statement like this: 'Fossil fuels are responsible for global warming'—which suggests moral responsibility but, in fact, states a causal link. For example, ClientEarth says on its website, 'Coal is the dirtiest of the fossil fuels and responsible for over 0.3°C of the 1°C increase in global average temperatures.'

To the extent that expert narratives circulate in public discourses, they are developed and deployed to influence audiences. Persuasion is the nature of the game, and narratives need to be intuitive, appealing, and easy to understand. Good narratives resonate with our cultural repertoire, they don't have to be explained or introduced. They are used strategically in public discourse to influence policymaking (Hajer 1995; Majone 1989; Riker 1986; Shiller 2017). They are never far away from 'noble lies,' statements that are made to influence the public and thereby bring about a desirable outcome. Some authors think manipulation of the audience is not a problem (see Banks 2021 for the example of climate change). As I will argue later on, we can distinguish two basic types of expert narrative which are important in this setting, alarming, and reassuring narratives.

The cases of climate change and Covid-19 fall squarely within the public policy narrative framework where narratives play a strategic role in the political debate. The chapters on the professions in this book show how private companies, practitioners, and social scientists are actively engaged in the construction of narratives. They also emphasize hope and fear, and thematize control and responsibility.

This chapter has outlined what I believe to be important differences between types of expertise in modern society. I have distinguished different social roles and functions that are important to consider. I have drawn attention to the rise of the commentator as a type of expert operating in public discourse and to the diffusion of systems of technological knowledge production, known as AI. Both of these will be discussed in this book. The simultaneous rise of both types poses a challenge to established expertise, as it is represented in official bodies, the professions, and science.

Notes

1 Scientists' strategies of boundary work have been examined by (Gieryn 1995). He identifies four types of boundary work: monopolization, expansion, expulsion, and protection.
2 This fallacy is very evident in Brennan's (2005) call for epistocracy.
3 The notion of lay expertise is mainly applied in research about medial settings, see, for example, Akrich and Rabeharisoa (2012). The *locus classicus* is (Epstein 1995:408), who claims that 'AIDS treatment activists have constituted themselves as credible participants in the process of knowledge construction, thereby bringing about changes in the epistemic practices of biomedical research.'

Chapter 2

Expertise in STS

Scholarship in Science and Technology Studies (STS) has brought forward a variety of contributions to the problem of expertise. The emphasis has been on the nature of scientific work, the relation between science and society, and the role of scientific advisory bodies as boundary organizations of expertise. These will be important reference points for the argument. However, a limitation of this scholarship is the tendency to use a rather narrow definition of experts as science based. Other relevant research in this field emphasizes the contribution made by laypeople, which has been conceptualized as 'lay' or citizen expertise. More generally, experts are located in the coproduction of knowledge and social order.

Interestingly, the songs *Hey Jude* and *Star-Spangled Bannner* both can be heard, although the former is more rarely played.

Laboratory studies

Harry Collins has made an important contribution for the analysis of expertise in contemporary society (Collins and Evans 2002, 2007; Collins 1985, 2014). Together with Rob Evans he critically reviews the move in science studies towards relativizing faith in science, and attendant calls for an extension of scientific expertise. They argue that there are good reasons to give scientific expertise a special place in society and its decision-making procedures:

> One of the most important contributions of the sociology of scientific knowledge (SSK) has been to make it much harder to make the claim: "Trust scientists because they have special access to the truth". Our question is: "If it is no longer clear that scientists and technologists have special access to the truth, why should their advice be specially valued?" This, we think, is the pressing intellectual problem of the age.
>
> (Collins and Evans 2002:236)

They pose a related question about the scope of non-scientists' participation in decision-making: 'How far should participation in technical

DOI: 10.4324/9781003320227-4

decision-making extend?' Science studies have shown that there is 'more to scientific and technical expertise than is encompassed in the work of formally accredited scientists and technologists, but it has not told us how much more' (Collins and Evans 2002:237).

There are two important issues here: one is about participation in decision-making, and the other is about participation in knowledge creation. Collins and Evans's term 'socio-technical decision-making' seems to relate to both. However, the authors do not elaborate on the potential ambiguity of this term. Instead, they introduce another distinction between interactional and contributory expertise. The first is defined as 'enough expertise to interact interestingly with participants,' and the second is 'enough expertise to contribute to the science.' According to Collins (2014:65), one becomes a contributory expert 'by working with other contributory experts and picking up their skills and techniques.' Interactional expertise, on the other hand, can be acquired by deep immersion in the linguistic discourse of the relevant scientific domain.

The distinction between interactional and contributory expertise makes sense if one wants to theorize the role of a scientific expert, more precisely, the role of an experimental scientist and Collins's own work has nearly exclusively focused on such cases. The notion of a core-set of scientists, which he introduced in his study of gravitational wave research (Collins 1985), he has later applied in research about expertise for policy decisions (Collins and Evans 2002). What counts for him are core-set competences that can only be acquired through practical participation on the laboratory bench. Only in this way can scientists gain the important tacit knowledge, and only in this way can scientists as experts claim to make competent statements.

However, the rise of knowledge workers in contemporary society points to the importance of 'interactional expertise,' which could lead to 'contributory expertise' through deep immersion in the discourse of the relevant domain.[1] The laboratory as the site of knowledge creation and the scientific institute which signals the competence of the researcher and thus makes her a 'certified expert' is not the only source of expert knowledge, and it is arguably not the most important one when it comes to political decision-making. Laboratories as sources of power have been posited by several authors conducting ethnographies of laboratories in the 1970s and 80s (Latour 1983). This special focus has perhaps led to an unwarranted generalization in terms of political theory.

Another aspect of the core-set terminology is worth noting. In his book *Changing Order*, Collins drew attention to the fact that core-sets are not only or mainly distinguished by competencies but by gate-keeping activities. Collins argued that social factors crucially decided what counted as scientific expertise, because there is no independent criterion available that could provide a standard for judging the competence of researchers and their knowledge claims at the frontier of new knowledge creation. Remember: *Changing Order* dealt with the problem of experimental replication and the attendant issue of experimenter's regress. In Collins's later work on expertise, the contributory quality of core-sets is distinguished from the interactive expertise of those outside the core-set. The separation is made on the basis of core competence, not on the basis of social mechanisms of gate-keeping. This losing sight of one of the distinctive features of his earlier work is noteworthy, and I will return to it.

In their book *Rethinking Expertise,* Collins and Evans state: '[W]e ought to prefer the judgments of those who "know what they are talking about."' In saying this, they recognize the role of non-certified and non-scientific voices. In their words, '[i]t opens to the door to Wynne's sheep farmers and to Epstein's AIDS sufferers and, perhaps, to anyone with interactional expertise . . .' (Collins and Evans 2007:113). This indicates that the authors are open to accommodating aspects of field expertise into the concept of expertise. In a later publication, Collins returns the issue, again trying to carve out a special place for scientific expertise. He thus ends his book with this plea: 'Science is inexact, messy and more often a matter of judgement than calculation. But if we want our judgements about the natural and social worlds made by good, disinterested people, then we should start the zeitgeist moving in the other direction and learn, once more, to elevate science to a special position in our society' (Collins 2014:132). This sounds very similar to the tune Naomi Oreskes has intonated, albeit a light version of *Hey Jude.*

Bruno Latour, like Collins, started his academic career doing laboratory studies, and, like Collins, he conceives of laboratories as sources of power. He does not deal with the role of the expert in society in a separate publication. In his oeuvre, he mentions the notion of expertise in passing, referring to the meaning of a specialist with experience, quoting the term *experitus.* For him, an expert is a specialist who has the luxury 'to work in the dark', where he can commit many errors, in contrast to the politician, who has to make all errors in public (Latour 1993:225). Latour's real attention is less on the role of experts, but on the role of science and scientists, on the role of laboratories as sources of political power, and on the similarity between science and politics.

Actor-networks, human and more-than human agency

However, Latour is most widely known for a bold theoretical move which is relevant for my argument. As Michael Lynch put it, 'In the history of the social sciences, there have been repeated efforts to explain social phenomena with natural scientific concepts and principles, often drawn from evolutionary biology and the neurosciences. Unlike such efforts to naturalize sociology, the constructivism that emerged from STS and cultural studies was a mirror image: a thoroughgoing enculturation and politicization of nature. Latour's (1987) Actor-Network Theory is perhaps the most persistent and prominent of these efforts in STS. That theory is, in part, a reaction to efforts to describe scientific developments sociologically, such as in Shapin's account of phrenology, or in Latour and Woolgar's earlier ethnography.' (Lynch 2016:108)

As Lynch notes, the second edition of their book *Laboratory Life* had deleted the word 'social' from its original sub-title (now reading *The Construction of Scientific Facts*), signalling the demise of any purely sociological account of scientific practice.

Those earlier accounts emphasized social interests, norms, and actions. In backgrounding the importance of 'the social,' Latour does not become a convert to naturalistic scientific claims on their own terms: 'Instead, he advises his readers to expand the domains of action and agency beyond the field of human actions, so that humans, collective agencies, non- human entities, theoretical causes, laboratory apparatus, and virtually all other nominal "things" placed in subject position in narratives about scientific and technological innovations, become "actors" (or "actants") that have "agency"' (Lynch 2016:107; see also Høstaker 2005).

His framework of actor-networks dispenses with explanation and only tries to describe networks, their reach, composition, and growth over time. Through this description, he wants to show and critically examine the genesis of established facts. However, established facts here pertain to a situation where despite uncertainty, and a competition of knowledge claims, a society comes to a conclusion of how to regulate a specific problem. This is why Latourian networks have no limits, and it does not make sense to speak about 'natural,' 'social,' 'political,' or 'scientific' factors. This flat ontology does not recognize the problem of political decision-making as a discrete issue and the role of expertise in the process.[2] It also is surprisingly silent on detailed questions of agency and the problem who can be held accountable or responsible for advice.

Werner Rammert has suggested a thought-provoking taxonomy of agency which advances our understanding in this regard. Rejecting the dichotomy between humans who have agency in the full sense (as meaning processing subjects with decision-making capabilities and moral responsibility) and machines which don't, he suggests to use a pragmatic definition of agency which sees agency in a gradual way. While the term 'gradual' seems counter-intuitive, there is an interesting lesson in his argument:

> This gradual and multi-level model of agency gives us the possibility to escape the dilemma of having to either reserve agency up to the humans or to flatten the concept of agency unnecessarily. Neither are we forced to claim that the activities of humans, machines and programs are substantially the same kind of behaviour. Nor do we have to stick to the conception that human action and technical operation are fundamentally different from one another.
>
> (Rammert 2008:77)

Using an example from civil aviation, which involves flying passengers to their holiday destination, he asks 'who is flying?,' i.e. who 'does' the flying? He lists all the necessary components to make this flight a successful operation.

These are the pilots, radio operators, flight controllers, the tourist office, the jet engine, radio equipment, autopilot software, radar, air traffic systems, and so on. Rammert (2008:85) describes these networked interactions between humans and non-humans as distributed agency and distinguishes between human action as 'distributed between many loci and instances that plan, control, and execute the activities.' It furthermore 'refers to hybrid constellations made of heterogeneous units of agency.' (ibid.). Finally, he adds a third step which is less frequently encountered as it goes beyond the traditional hierarchical model of instrumental logic, in which technology is very much seen as in a master-slave relation. In this third step, '[t]he technological units are given more freedom of choice and higher levels of agency in order to enrich their capacity of assistance and to strengthen their role as relatively autonomous agents.' (ibid.).

His comment about the future role of such systems was prescient: 'Though the mode of framed interactivity has rarely been implemented up to now, this mode may become a new paradigm for the design of future constellations. It is currently sought after in many different places: in laboratories of distributed intelligence, in research and development clusters on robotics, man-machine interfaces and new media design, as well as in the studios of interactive artists, in the media labs of the entertainment industry, and at the software benches of videogame developers' (Rammert 2008:86).

Interestingly, he does not list the cases of journalism, law, or medicine. Following Giddens, Rammert distinguishes between three different levels of agency: causality, contingency, and intentionality (here, I will limit myself to the first two and discuss intentionality later). He shows convincingly that there is little difference between human and non-human agency with regard to causality. In both cases, an intervention is performed, and as a result, the state of the world has changed. With regard to contingency, what matters is the capacity to act in a different way and to choose between options. Where technologies perform in this way, they no longer follow the 'paradigm of command and execution: Instrumentality is replaced by relations of interactivity.' Rammert uses the example of interactive videogames, which create spaces of high virtual contingency that simulate human user's action. These technologies function like a Turing test: they make it nearly impossible to discriminate between human-enacted and computer enacted characters in the play' (Rammert 2008:76).

The aspect of contingency raises a question which was essential to classical social theory: how can social order emerge under conditions of double contingency between Ego and Alter? As we know, the solution to this problem was seen in social norms and regulations, which provide reciproocity, reliability and stability of expectations (Vanderstraeten 2002). The Legal theorist Gunther Teubner has argued that this provides the model for legal reasoning, which was also applied to legal persons (aka organizations or corporate actors), a decisive innovation in law and society. He also thinks that non-human agency can be subsumed under this logic via analogy. Hybrids, networks comprising human and non-human elements, are linked to social order through meaning-processing humans within these networks. Teubner stresses the fact that these networks are something different from corporate actors because the non-human elements make their own distinctive contributions. Thus, 'the hybrid itself develops its own phenomenal world view, its self-perception as a living entity, its own order of preferences, its own social needs and political interests, different from an individual or collective acting on its own' (Teubner 2006:512).

Teubner concludes: 'The result of all this is that indeed non-humans gain access to social communication, albeit in a rather indirect way. The law plays a special role in this game; it stabilizes non-human personality by granting legal status to the hybrids via the construct of the juridical person, by attributing to them the capacity to act, by giving them rights, burdening them with duties, and making them liable in several forms of legal responsibility' (Teubner 2006:515). This is a pertinent reminder that exposes a blind spot of actor-network theory. In its sweeping enthusiasm for making a case for non-human agency, it went too far and forgot some powerful social mechanisms. It is, after all, the state and law who determine what is legal, and how interests and needs are represented in society.[3] We, therefore, need to look at the practice of the legal system and how it deals with the rise of non-human agency. The hard case is artificial intelligence to which I will return below.

This short review of laboratory studies indicates that they are a mixed blessing for the ability of the field of STS when tackling the problem of expertise in society. They have focused on scientists and their practices and their involvement in a wider net of relations. Political and institutional analysis has been a weak point in this field. In what follows, I will look at contributions from authors that have a different theoretical agenda, focusing on the question of decision-making in modern democracies where expert knowledge is seen as vitally important.

Regulatory science

In their study of advisory committees, Salter et al. (1988) persuasively argued that there is a significant difference between science and scientific research on the one hand, and knowledge that is applied to solve policy issues on the other. They called the latter mandated science in order to draw attention to one type of science that was not the outcome of an autonomous research process but commissioned by public agencies keen to get specific and practical advice on regulatory policy issues.

As the authors put it, 'Our image of scientists pictures them at work in the laboratory; we seldom raise the question of how scientific information moves from the laboratory to the world of politics and policy making' (Salter et al. 1988:1).

The work of such committees uses all the terms we are used to hear about scientific research, such as literature review or peer review, but the purpose of mandated science is not to produce new scientific findings. Its point is to make a judgement about multiple sources of evidence, resulting in a recommendation to a pressing problem of public policy.

However, as the authors argue, this task is difficult, if not impossible to solve:

> Mandated science must be understood as a separate sphere of scientific work... Increasingly, decision makers and their publics are placed in a quandary. On one hand they are increasingly dependent on science and scientists... on the other hand it is increasingly apparent that science cannot provide the clear answers that governments seek, at least not at the time when regulatory decisions are required. Moreover, science often provides conflicting answers ...
>
> (Salter et al. 1988:4)[4]

While the book by Salter et al. received little attention, Sheila Jasanoff's work on advisory committees was to become highly visible, mainly through her book *The Fifth Branch,* in which she addresses the issue of expertise using the term regulatory science. In the opening chapter, she lists three major findings from the sociology of science 'that must be taken into account in any serious discussion of scientific advising' (Jasanoff 1990:12). These are: (1) Scientific facts are socially constructed; (2) Scientific paradigms and social prestige are important for problems facing advisory committees; and (3) Through boundary work, scientists decide who belongs to relevant professional and policy communities, thus holding up an appearance of scientific authority even in the face of uncertainty. Jasanoff draws an explicit comparison between conventional science and regulatory science. In so doing, she addresses a shortcoming of Collins's work (mentioned above). Experts working for the regulatory state are not only or mainly in their position as experts because of their technical competence but because of mechanisms of social inclusion/exclusion.

Jasanoff contrasts a technocratic with a democratic model of science advice (see also Habermas 1971b; Irwin 1995) and concludes that neither 'captures accurately what is at stake in decisions that are at once scientific and political. The notion that the scientific component of decision-making can be separated from the political and entrusted to independent experts has effectively been dismantled by recent contributions of the political and social studies of science.' And, even more strongly: 'The idea that scientists can speak truth to power in a value-free manner has emerged as a myth without correlates in reality' (Jasanoff 1990:17, echoing Collingridge and Reeve 1986).

Outlining the structure and rationale of the argument of her book, Jasanoff emphasizes the central role of scholarship from the sociology of science beyond the 'expected' fields of law, political science, and policy analysis. While the central role of knowledge is highlighted, the focus on its scientific nature posits a problematic and perhaps unwarranted assumption.

Writing on experts as policy advisors Jasanoff (1990:229) points out that 'experts themselves seem at times painfully aware that what they are doing is not "science" in any ordinary sense, but a hybrid activity that combines elements of scientific evidence and reasoning with large doses of social and political judgment.' While this statement captures an important element, it also shows that the role of expert is measured against the role of the scientist. This background assumption seems to be made in several contributions about expertise.

Jasanoff (2011b:21) also emphasizes the role of the expert as a translator or mediator between knowledge and decision-making, quite akin to what I propose. However, her concept of the expert seems to depict them as professionals: 'it is not science per se that speaks unmediated to power, but [...] the bridge between science and politics is built by experts, a cadre of knowledgeable professionals.' Occluded from this conceptualization is the notion of expertise that is not based on esoteric science and not located within a profession.

This section provides evidence for the presence of a tune which does not take official claims of experts on face value but problematizes expertise. However, we have also seen that there is a temptation to conceptualize experts as either scientists or professionals. It is now time to consider the case of citizen or lay expertise.

Citizen expertise

Over the past decades, modern societies have experienced a 'significant increase in organized challenges to policy domains that are highly scientific and technical in character,' as Parthasarathy (2010:355) put it. Laypeople or members of the public who are affected by policies and decisions are demanding a say, and a place at the table. However, they face obstacles when launching their challenges. There are formal and informal rules of policy domains which can make it difficult for those without technical expertise to engage as equals (Irwin 1995).

Throughout this book, I will refer to this expertise 'from below' as citizen expertise, field expertise, or lay expertise (for an excellent discussion, see Krick 2021). These terms are used synonymously. It is important to note that these experts 'from below' base their expertise on various elements. Some are based on experience of, and personal interest in an issue, others are based on cognitive knowledge claims, and there are mixtures between them. If one were to draw up ideal types, imagine one where people are expressing their discontent with an official policy, which uses knowledge claims to legitimize the policy. In this case, the knowledge claims would not necessarily be contested, or only in a very general way. Another type is engaging with the knowledge claims in detail, for example, by reading scientific reports and commenting on the content in a critical way. To take two influential exemplar cases, Brian Wynne's Cumbrian sheep farmers after the radioactive fallout from the Chernobyl nuclear accident falls into the first category, and Steve Epstein's AIDS activists fall into the second.

Wynne calls for an inclusion of laypeople in decision-making. His critical account of the arrogance of state authority has proven very influential with scholars in STS and deliberative democracy circles. Wynne shows how government scientists provided laboratory and abstract expertise that first gave false reassurances and then imposed restrictions which affected the farmers' economic interests. But there is a question to be asked about which social groups represented which expertise. It is not clear what difference in policy recommendations could have come from the farmers' lay expertise, apart from the obvious economic aspect of financial compensation for the economic loss.

Wynne does provide some hints about the economic dimensions of the decisions at stake. Hill farmers were dependent on their lambs, which were raised after Easter and sold in the Autumn, mainly to European markets. They needed to be sold at the right time, being neither too lean nor too fat. Any indication or suspicion that the lambs were radioactively contaminated would massively devalue them and thus pose an immediate and direct threat to the farmers' livelihoods. What was the right thing to do given the circumstances of radioactive contamination of soil and herds? This is a question that is never explicitly addressed in Wynne's account. We hear about the distrust of local farmers vis-à-vis the government scientists and a history of local radioactive pollution following the Windscale disaster in 1957, which was probably covered up through government secrecy. But given the circumstances of 1986, what were the options for the farmers and for the government? What was in the public interest, and what in the farmers' interests? Was there a difference between the two? How did the expertise of the government scientists and that of the Cumbrian farmers answer these questions? While Wynne's account tells the story of an apparently confused and haphazard government intervention it does not tell us much about the farmers' demands with regard to managing the crisis.

For example, Wynne relates that 'Although the farmers accepted the need for restrictions, they could not accept the experts' apparent ignorance of the effects of their approach on the normally flexible and informal system of hill farm management. This experience of expert knowledge being out of touch with practical reality and thus of no validity was often repeated with diverse concrete illustrations in interviews' (Wynne 1989:34). In addition, conflicts about the rules for compensation emerged, which left the farmers embittered.

The culture and social identity of the hill farmers are mentioned as important factors when assessing the divergent ways of dealing with the risk of radioactive fallout on the Cumbrian hills. As Wynne points out, the farmers' expertise was not codified anywhere, it was passed down the generations orally and by apprenticeship.

In sum, then, Wynne's point is not that the hill farmers have developed some kind of counter-expertise that would have led to different policy recommendations. Rather, he is concerned about the interests and social identity of the hill farming community and about the lack of humility on the part of government experts. The hill farmers, for their part, were much more aware of the limitations of knowledge when trying to evaluate the situation and devising a course of action.

Epstein shows how patient groups and activists exerted influence at the base of political decision-making regarding AIDS treatment in the USA, even changing the practice of clinical studies and scientific research. According to Epstein, it would make little sense to speak of an extension of the realm of scientific practice in order to integrate the viewpoint of laypersons. Epstein has no such science-centred viewpoint. Rather, he emphasizes that the pressure exerted by laypersons proved to be central and cannot be seen as only an additional contribution to expert discourse.

He writes about the activists:

> Their publications have created new pathways for the dissemination of medical information. Their pressure has caused the prestigious journals to release findings faster to the press. Their voice and vote on review committees have helped determine which studies receive funding. Their efforts have led to changes in the very definition of 'AIDS' to incorporate the HIV-related conditions that affect women. Their interventions have led to the establishment of new mechanisms for regulating drugs, such as expanded access and accelerated approval. Their arguments have brought about shifts in the balance of power between competing visions of how clinical trials should be conducted. Their close scrutiny has encouraged basic scientists to move compounds more rapidly into clinical trials. And their networking has brought different communities of scientists into cooperative relationships with one another, thereby changing patterns of informal communication within science.
>
> (Epstein 1996:338–339)

These ideal types of citizen expertise, experience based, and engaging with the science certainly overlap. There is no interest-free expertise or expertise without experience. However, the degree to which citizen experts engage with esoteric knowledge claims varies considerably.

Michel Callon (1999) has taken up some of these concerns, especially the point about government scientists patronizing citizens, and citizen experts. The latter fear 'that someone else may decide for them what is good for them, and that such decisions would be taken without the slightest knowledge of their needs or wishes' (Callon 1999:88; see also Roszak 1968).

This is an important insight which resonates with concerns voiced by other scholars from the field of economics and psychology (see Chapters 4 and 6). It underlines the fact that different forms of expertise may be competing and that expertise may be closely linked to a social interest. Callon mentions those affected by official expertise, a scientific or technocratic undertaking that is perceived as threatening by those affected. One could argue that these scientific experts themselves are trying to protect their interests, too. But there is a difference in that the lay experts often cannot act in a symmetrical way, and thus they do not have the same kind of control over events.

Callon's approach is quite close to what I am proposing here. In this view, both lay experts and official experts mediate between a body of knowledge and decision contexts. Both fight for recognition and acceptance (albeit the official expert is likely to have an advantage in this game, especially under conditions where access to power elites is important). The nature of knowledge in both cases is not essential to the framework: it could be a body of scientific knowledge, it could be a body of practical knowledge, of tacit, or of secret knowledge (see Stehr and Grundmann 2012).[5] The important aspect is that expertise mediates between a body of knowledge and its application. Scientists could fulfil this role, but they are not the only group.

Wynne, Epstein, and Callon all have developed the notion of stakeholder inclusion during controversial public debates, which becomes even more prominent in the framework of Postnormal Science, a situation in which '[s]takes are high, facts are uncertain, values in dispute, and decisions urgent' (Funtowicz and Ravetz 1990; Ravetz 1999). In postnormal situations, science and decision-making become closely coupled. Does this bridge the gap between science and politics? Or is this entanglement a hindrance for decision-making? The authors believe that credible solutions must rely on the legitimation of public participation. Science should stop pretending that it can provide reliable knowledge for decisions. Under conditions of uncertainty and value conflict, experts and laypeople are alike and should be given equal treatment in the process:

> Persons directly affected by an environmental problem will have a keener awareness of its symptoms, and a more pressing concern with the quality of official reassurances, than those in any other role. Thus they perform a function analogous to that of professional colleagues in the peer review or refereeing process in traditional science, which otherwise might not occur in these new contexts.
>
> (Ravetz 1999:649)

In a similar way to Callon and Wynne, Ravetz says that official expertise and lay expertise should be seen as complementary, not as antagonistic:

> The Post-Normal Science approach should not be interpreted as an attack on the accredited experts, but rather as assistance. The world of "normal science" in which they were trained has its place in any scientific study of the environment, but it needs to be supplemented by awareness of the 'post-normal' nature of the problems we now confront.
>
> (Ravetz 1999:653)

Here the stakeholder engagement is modelled explicitly on the institution of peer re as practiced within scientific communities. It is not entirely clear what the reason for inclusion is: is it the lay knowledge of stakeholders, or is it the democratic principle of inclusion of stakeholders, irrespective of their status as experts? If laypeople have special knowledge that is thought to be essential, or at least an important complement to the specialist knowledge, this needs to be spelled out. Alternatively, laypeople make judgements in the face of uncertainty, like experts do. Because they are similar in this regard, one cannot justify their exclusion from decision-making.

Yearley (2000:109) interprets the conundrum that Post-Normal Science (PNS) is more concerned with the quality of knowledge than with democratic inclusion as follows: 'Funtowicz and Ravetz stress that citizen involvement is proposed by them not because they are committed to the furthest extension of democracy, but because the involvement of a larger group of peers, with different kinds of knowledge, will be beneficial to the production of high-quality knowledge.' I will return to the virtue of diversity within expert groups below.

Others have warned not to exaggerate the differences between normal and post-normal science:

> However, in haste to acknowledge the political, there is a danger of setting up a dichotomy between times when issues were tame and science was normal, and now, when issues are wicked and a post-normal approach is required, perhaps more deeply focused on social discussion and learning. Did such a change ever really happen; has science ever been normal, or is it more that the political conflicts have now moved into a different, more public, arena?' (Turnpenny et al., 2011:301).

This goes back to the question from above about the opening up of apolitical issues that have been dealt with by technical specialists. However, this is issue dependent, and should not be seen as an expression of post-normal conditions.

In contrast to science-based notions of expertise, the contributions in this section emphasize lay expertise and lay-expert interaction. However, they stop short of conceptualizing the fundamental difference between knowledge production and decision contexts, and the attendant question of the specific properties of the *type of knowledge* that can be used for decision purposes. It leaves out the possibility of experts mediating between knowledge production and application.

Honest brokers

A conceptual refinement of the role of the scientist as expert has been developed by Roger Pielke Jnr. (2007). His typology identifies four different roles of scientists engaged in different ways with a decision-making process. These roles are presented as pure scientist, science arbiter, issue advocate, and honest broker. He calls them experts, but on several occasions, he also uses the term 'scientists and other experts' (without defining what these 'other experts' might be).

The pure scientist has no interest in the decision-making process and simply wants to share information about facts. 'The Science Arbiter serves as a resource for the decision-maker, standing ready to answer factual questions that the decision-maker thinks are relevant. The Science Arbiter does not tell the decision-maker what he or she ought to prefer' (Pielke Jnr. 2007:2). In contrast, the issue advocate tries to convince the decision-maker of one best course of action. Finally, the honest broker leaves it to the decision-maker to reduce the options and to make a choice: 'The defining characteristic of the honest broker of policy alternatives is an effort to expand (or at least clarify) the scope of choice for decision-making in a way that allows for the decision-maker to reduce choice based on his or her own preferences and values' (Pielke Jnr. 2007:2–3).

A characteristic of both honest brokers and issue advocates is an explicit engagement with decision alternatives, whereas the pure scientist and science arbiter are not concerned with a specific decision but instead serve as information resources. Unlike the science arbiter, the honest broker seeks explicitly to integrate scientific knowledge with stakeholder concerns in the form of alternative possible courses of action.

Pielke argues that 'contemporary science policies create strong incentives for scientists to wage political battles through science by emphasizing the roles of Pure Scientist and Science Arbiter in all cases, when in fact an advocacy position is actually being expressed. At the core of the argument presented here is a critique of the longstanding expectation that science can and should be separated from considerations of the applications of science' (Pielke 2007:77).

Pielke follows E. E. Schattschneider when saying that

> democracy is a competitive system in which the public is allowed to participate by voicing their views on alternatives presented to them in the political process. Such alternatives do not come up from the grassroots any more than you or I telling an auto mechanic what the options are for fixing a broken car. Policy alternatives come from experts. It is the role of experts in such a system to clarify the implications of their knowledge for action and to provide such implications in the form of policy alternatives to decision-makers who can then decide among different possible courses of action.
>
> (Pielke Jnr. 2007:12).

Two aspects of honest brokering are relevant. The first one takes up an insight by Schattschneider about the role of power in politics. The second is about the semantics of the terms 'honest' and 'brokering.' Schattschneider made the legendary statement that 'The definition of the alternatives is the supreme instrument of power … He who determines what politics is runs the country, because the definition of alternatives is the choice of conflicts, and the choice of conflicts allocates power' (Schattschneider 1960:68).

Tasking honest brokers with the identification of policy alternatives potentially assigns them a role of power. This could lead to an interference with politicians who usually want to keep the prerogative of defining policy options, and align their advisory resources accordingly. In fact, one may arrive at the conclusion that whenever experts try to monopolize a problem definition and its potential solutions, politicians are getting nervous. The case of climate change is maybe the most instructive example, as I will show in Chapter 7.

Secondly, both terms 'honest' and 'brokers' are problematic. What counts as honest could be a difficult judgement, depending on the merits of each case, to evaluate the honesty of such brokering. The main problem with the term, however, is the suggestion that experts as experts could stay aloof from the decision process which they have been asked to join. Brokers make matches, select options, and suggest courses of action. In this sense, experts are brokers, and it would be odd to restrict their role to a widening of policy options (which Pielke says is their 'defining characteristic'). Maybe Pielke does not see this as a problem because he seems to hold on to the ideal of impartiality as a virtue of experts.

The upshot of this section is that experts are seen either in close analogy to a scientific ideal, neglecting the role of non-scientific experts. Even in cases where issues of political participation and democratic inclusion are raised (as in PNS), the process is modelled on scientific practices (extended peer review). Callon and Wynne put more emphasis on lay expertise and highlight the need for mutual dialogue and inclusion in decision-making processes. Pielke's notion of expertise is rather science-centred, and while distinguishing important different roles of scientific advisors, he depicts honest brokers in a way which seems too limiting in my view.

Notes

1 It is surprising that Collins does not think that deep immersion for an extended period of time would allow participants to develop contributory expertise. Maybe he overgeneralizes his own experience when he says that decades of close contact to the community of gravitational wave researchers have not allowed him to become a contributor, even after he successfully accomplished something like a Turing test (Collins 2014:69–71).

2 However, in a recent talk about climate change, Bruno Latour (2013) does consider the nature of politics, which he defines as an agonistic field, following Carl Schmitt. Perhaps the turn from 'matters of fact' to 'matters of concern' (see Latour 2004) has led him to embrace a framework in which political aspects become important and can be addressed as such.

3 Helga Nowotny recounts the short story by Ian McEwan ('Düssel …') in which the first android becomes pregnant, posing important questions about human nature. She concludes, 'After all those turbulent years of anguished reflection and passion, the law had decided what it meant to be human' (Nowotny 2021).

4 See also Weinberg (1972:209), who famously said that there are 'questions which can be asked of science yet cannot be answered by science.'

5 The asymmetry between corporate and government expertise on the one hand and lay expertise on the other is perhaps most pronounced in the area of surveillance, as Edward Snowden's revelations have shown.

Chapter 3

The Power of the Professions

Insiders and outsiders

In their classic book *The Social Construction of Reality*, Berger and Luckmann describe the development of professional knowledge specialisms in modern society, which they call 'sub-universes.' They explore the consequences of their development and autonomy for society:

> The increasing number and complexity of sub-universes make them increasingly inaccessible to outsiders. They become esoteric enclaves, "hermetically sealed" ... to all but those who have been properly initiated into their mysteries. The increasing autonomy of sub-universes makes for special problems of legitimation vis-a-vis both outsiders and insiders.
>
> (Berger and Luckmann 1966:104)

In the following quote, they describe the logic that follows from this separation between insiders and outsiders in a general way before applying it to the case of medicine. The general description goes as follows:

> The outsiders have to be kept out, sometimes even kept ignorant of the existence of the sub-universe. If, however, they are not so ignorant, and if the sub-universe requires various special privileges and recognitions from the larger society, there is the problem of keeping out the outsiders and at the same time having them acknowledge the legitimacy of this procedure. This is done through various techniques of intimidation, rational and irrational propaganda (appealing to the outsiders' interests and to their emotions), mystification and, generally, the manipulation of prestige symbols. The insiders, on the other hand, have to be kept in. This requires the development of both practical and theoretical procedures by which the temptation to escape from the sub-universe can be checked.
>
> (Berger and Luckmann 1966:105)

In terms of musical tunes, this is one which has probably the most simple and recognizable melody from the choir of those critical of expert authority.

DOI: 10.4324/9781003320227-5

Berger and Luckman have given a vivid description that applies to many examples we can marshal in our imagination. Here is an example of the esoteric sub-universe of medicine: 'The lay public must be convinced that this is right and beneficial, and the medical fraternity must be held to the standards of the sub-universe. Thus the general population is intimidated by images of the physical doom that follows "going against doctor's advice"; it is persuaded not to do so by the pragmatic benefits of compliance, and by its own horror of illness and death' (Berger and Luckmann 1966:105).

They describe an archetypical narrative about the consequences of going against expert advice. Because experts know so much more than the ordinary citizen, the latter don't know what kind of fate is awaiting them if they go against the experts' counsel. The fear mongering rhetoric is needed as the patient's consent is required in most circumstances.

However, implied in the above narrative is its opposite: expert narratives not only have the element of fear but also of reassurance. Catastrophe can be averted if people follow the experts' counsel. This is not stated explicitly by the authors, but I think it follows from the logic of their argument. My working hypothesis is, therefore, that a typical expert narrative in areas where potential danger looms is either raising fear, or reassuring the audience, or doing both.

Berger and Luckman go on to specify how professional authority is constructed in medicine:

> To underline its authority the medical profession shrouds itself in the age-old symbols of power and mystery, from outlandish costume to incomprehensible language, all of which, of course, are legitimated to the public and to itself in pragmatic terms. Meanwhile the fully accredited inhabitants of the medical world are kept from 'quackery' (that is, from stepping outside the medical sub-universe in thought or action) not only by the powerful external controls available to the profession, but by a whole body of professional knowledge that offers them 'scientific proof' of the folly and even wickedness of such deviance. In other words, an entire legitimating machinery is at work so that laymen will remain laymen, and doctors doctors, and (if at all possible) that both will do so happily.
>
> (Berger and Luckmann 1966:105)[1]

Such strategies are important, especially in situations where experts as commentators operate in public debates where they want to distinguish themselves from voices that they perceive as ignorant, interested, or in other ways illegitimate. However, this very attempt is likely to fuel further suspicion and debate.

Others have described the role of the professions in modern society in similar terms. In his book *The System of Professions*, Andrew Abbott surveys the literature on the professions and on the process of professionalization. For him, the terms occupation and expertise are closely related to the professions and he thinks that the key issue here is that occupational groups are trying to control knowledge and skill. In a quasi-Foucauldian perspective, he sees the constant struggle between different occupational groups as the driving force. According to Abbott, it does not make sense to try and provide ideal typical definitions of professions and of the process of professionalization because these change over time and across countries. Nevertheless, he does provide a very general definition of a profession as 'exclusive occupational groups applying somewhat abstract knowledge to particular cases' (Abbott 1988:7).

Owning the problem

Abbot starts with this question of occupational control and points out that it can be achieved in two ways, either by some group directly controlling its technique or by defining who may be counted in. 'The other form of control involves abstract knowledge. Here, practical skills grow out of an abstract system of knowledge, and control of the occupation lies in control of the abstractions that generate the practical techniques' (Abbott 1988:8). Abbott thinks only the second, the aspect of abstraction, can be used for distinguishing a profession, allowing it to win an interprofessional competition, and to 'defend it from interlopers, and seize new problems' (Abbott 1988:9). Who owns a problem is a constant issue in these debates. Abbott proposes the term jurisdiction, which is supposed to show the link between a profession and its work (Abbott 1988:20).

Gieryn (1995) applies the terms formalization and codification as instances of abstraction and thinks that for a profession to be successful in establishing its independence its needs to strike a delicate balance between the two: 'Too much formalization or codification of expert knowledge makes tasks vulnerable to routinization, and they may be easily usurped by less expensive paraprofessionals (or delegated to subordinates). At the other extreme, too little formalization makes its jurisdiction vulnerable to technological or other exogenous changes.' This insight resonates with my conceptual scheme of citizen experts and AI systems invading the space of a profession.

Abbott's account can be sketched briefly as follows. The professions crucially rely on expertise, as only expertise can help us solve problems. Because expertise is located in the professions, only professionals can help solving problems. By linking problems, experts, and professions in this way, there is by definition no role left for non-professional experts. Using the example of alcoholism in the USA, Abbott says that it was first seen as a moral problem, so the church ministers got involved by way of condemnation and forgiveness. Then it was seen as a medical problem with the promise of cure, as the doctors took over. It then became redefined as a legal problem, and people were sent to prison. Psychiatrists defined the problem as one of mental disturbance. But when the problem was defined as a political problem, experts had lost their professional power of expertise.

This sketch reveals two important limitations in Abbott's theory. He seems to postulate that there is always one profession in possession of the power of expertise. Secondly, the lay public is excluded from any role in problem-solving, but as a matter of fact, laypeople may become experts while not being part of a profession. If we instead allow for the co-presence and influence of several professional groups, and interest groups, the picture changes, and we realize that the solving of problems in society cannot be reduced to technical problem-solving. At the same time, when a problem makes it onto the agenda of public policy, this does not mean that experts have lost their influence.

Networks of expertise

Gil Ayal takes up some of these concerns in his attempt to explain the rise of autism as a social problem. His analysis shifts the attention away from single experts and professions to the distributed network of expertise. He shows how an alternative 'network of expertise' could arise in which the key actors were no longer child psychiatrists, as had been the case with professional experts such as Leo Kanner:

> The key actors in assembling this network were not child psychiatrists, but the parents of children with autism. Seeking to modify the relations within Kanner's network of clinical expertise, which expropriated their knowledge and stigmatized them as "emotional refrigerators," they set up alternative mechanisms of data collection and experimentation, which bypassed the clinician. To do so, they allied themselves with behavioural psychologists and occupational therapists who, due to their peripheral position, developed therapies that blurred the distinction between mental illness and retardation as well as the distinction between lay and expert. It was this new actor-network, composed of arrangements that blurred the boundaries between parents, researchers, therapists, and activists, that was finally able to "solve" the problem. The combination of this new actor, the new ecology created by deinstitutionalization, and the capacity of the therapies to secure the cooperation of the patients themselves, is what led to the autism "epidemic"'.
>
> (Eyal 2013:868)

In parallel to the alcoholism example given by Abbott, the influx of laypeople into the problem domain led to a reconfiguration of expertise. But this does not mean that professional experts ceased to play a role, although they may have lost the ownership of the problem.

Eyal (2013:864) is not content with the view that social problems arise because they have been socially constructed and are a product of attribution. Drawing on the work by Hilgartner and Bosk, he states: 'Something is a social problem, … not because it is an objective and identifiable condition that has intrinsically harmful effects but because such effects have been collectively attributed to it. … the approach developed [here], focusing on the arrangements and conditions necessary for problems to become objects of expert labor, is geared precisely to overcome this opposition between what is real/objective and what is merely attributed/socially constructed.'

Eyal sees the solution to this false dichotomy in the notion of diffusion of expertise through society. He makes the point, quite convincingly, that we should not only understand the professions' desire for monopoly and autonomy but the spreading of their knowledge base to wider audiences. Individual experts may lose influence, but the body of knowledge, their expertise, gain wider acceptance. Commenting on Epstein (1995) and his work on AIDS activists becoming part of the expert community, he states:

> Could we not say, then, that even as the power of the experts, the monopoly and autonomy of medical doctors, was diminished, the power of medical expertise was in fact augmented by the activists because now "people with AIDS and HIV infection would willingly participate in the trials and conform to the protocols," that is, now there was a mechanism of exchange that secured patients' (and experimental subjects') cooperation? Could we not say the same about the recent trend in self-diagnosis, greatly assisted by the Internet and by the spread of evidence-based medicine? Doctors may lose jurisdiction, but the network of medical expertise is extended via generosity and dialogue.
>
> (Eyal 2013:876)

Eyal's argument is based on the possibility to draw a distinction between experts and expertise. The former are persons based in professional life, and the latter are networks of expert knowledge, including laypeople, artefacts, and technologies. Eyal puts it this way:

> This approach combines the considerable achievements of jurisdictional analysis with an analysis of how forms of expertise are gradually assembled. It differs from the sociology of professions by being inclusive of all who can make viable claims to expertise.
>
> (Eyal 2013:899)

I have two reservations about this argument. The first is that it loses sight of the problem noted above, where (professional) experts try to monopolize public discourse. By extending the notion of expertise, we gain a wider analytical notion but lose the critical angle on experts' rhetorical strategies and self-interest.

The second reservation relates to the inclusion of laypersons into networks of expertise. While this is a convincing point, it stops short of conceptualizing laypersons as experts themselves. Only on occasions does Eyal say that parents can be experts regarding their child's well-being, which means that laypeople, who by definition are not experts, can only participate in the game via the contribution to expertise, as a network and body of knowledge. He wants to carefully distinguish 'between experts and expertise as two modes of analysis that are not reducible to one another; by asking not only who controls a task and how jurisdictional boundaries are assembled but also what arrangements, devices, concepts, and other actors are necessary if an expert statement or performance is to be formulated, reproduced, and disseminated as an immutable and combinable mobile' (Eyal 2013:899).

Coming back to the musical metaphor, it seems that the literature on the professions, especially in Abbott's version, tends to favour the simple riff where one can define and distinguish real expertise more or less clearly. However, the coda about fake experts and their motivations is largely absent.

Tacit knowledge?

This raises an important question about the embodiment of expert knowledge. Is it tied up with persons and their personal skills and experience? Or can it be understood as distributed more widely, which can be embedded in persons, organizations, and technology?

Science and Technology Studies (STS) scholars have different answers to this question, depending on who you ask. Cambrosio et al. (1992:355) argue that expertise is an emerging network property. The construction of expert knowledge goes hand in hand with the construction of a network which establishes formal or informal standards: 'Standards will then function as black-boxes, which will possibly be used in the construction of further, more extended networks.' The authors argue further that expertise can be best understood as an 'ongoing accomplishment, rather than as the unquestioned mobilization of well-defined, pre-existing bodies of knowledge.' Eyal is arguing along these lines as well.

In contrast, Harry Collins holds the view that expertise presupposes a real and substantive skill possessed by an individual, typically by virtue of being socialized into a group of similar experts. He distinguishes between contributory expertise and interactional expertise. What differentiates the true expert from the merely competent, or contributory expertise from interactional expertise, is the embodied and tacit mastery of this background of practices, the 'form of life' shared by the 'core set.'

Eyal objects to this on two grounds. Firstly, he thinks that there is a lurking normativism and the assumption that sociologists could identify the two forms of expertise and thus act as an arbiter in these matters. There is the additional assumption that one of these forms of expertise is superior or more reliable than the other. Collins has taken this step by commenting on issues such as vaccination and climate change (Collins 2014). I share Eyal's critique:

> Collins and Evans embrace this normative role and claim that in a world of vaccine scares and global warming controversies it is indispensable that someone play this role. I have severe doubts, however, that sociologists can play this role without themselves becoming embroiled in a controversy about their own expertise and credentials.
>
> (Eyal 2013:870)

Here we hear the dissonant tunes from within STS. On the one hand, a tune like Collins', who thinks there is a case to be made for distinguishing real expertise from pseudo-expertise. On the other hand, a tune like Eyal's, who thinks such a neat separation is not possible, at least not possible without compromising one's own position as observer of the expertise in question.

The second point Eyal raises against Collins relates to the tacit skills of persons. He observes that 'if we want to account for the superior and speedy execution of a task it is not enough to focus on the actors and their skills' (Eyal 2013:871). His point is that we also need to include the tools and devices used in the performance of the task and institutions such as regulatory agencies and standards. Echoing Cambrosio et al., he cautions that we should also include the role played by technological systems and the makers of these devices. 'This complex make-up of expertise is typically much more evident when it is still "in the making" and alternative devices, actors, concepts, and arrangements are still viable candidates for formulating the problem or addressing it. Once a form of expertise has been developed, however, this complex make-up is obscured by a process of "black boxing" and standardizing that makes it appear as if expertise is embodied by the expert' (Eyal 2013:871).

In this process of instituting expertise (as a system), the expert (as a person) emerges and is then attributed with the autonomy and power which is said to reside in her. Expertise is embodied in a person, obliterating the role of its genesis. A classical process of inversion of reality seems to have taken place (Latour and Woolgar 1986, 177; Marx 1975).

This raises two related and important questions. Do we see a movement from person-based expertise to system-based expertise? And is there a shift in trust from trust in persons to trust in systems (see, e.g. Luhmann 2020 and Nowotny 2021)? I will return to these questions in the Afterword.

The next chapter will deal with a specific example of embodied expertise, persons who allegedly possess special skills that make them very good forecasters. Their personal traits and forms of cooperation are at the centre of this analysis.

Note

1 Berger and Luckmann express a type of scepticism towards experts, typical of the time. Theodore Roszak's *The Making of a Counterculture* appeared in 1968, followed by others in the early 1970s, such as EF Schumacher's *Small Is Beautiful,* or Ivan Illich's *Tools for Conviviality* (see below). Berger and Luckmann's book has been cited over 70,000 times, according to Google Scholar. Given this towering contribution, it is odd that some of the literature referenced here seems to be oblivious to it. Neither has it been cited by some prominent STS scholars, nor by theorists of the professions such as Andrew Abbott. Even a philosopher like John Searle, who wrote a book with a very similar title in 1995, did not engage with Berger and Luckmann's contribution (Lynch 2016).

Chapter 4

Predicting the Future

It is no surprise that predictions and forecasts are relevant to the topic of expertise. Recommending courses of actions presupposes some understanding about likely developments based on causal assumptions about how the world works and how it is going to develop. Forecasters are examples of experts which offer predictions, even if they do not directly or explicitly offer advice. Clients can be presumed to be interested in those experts that are known for reliable predictions in the past. However, much still depends on the institutional settings and incentives. Even if proven superforecasters are available on the market of experts, it does not follow that their services will be sought, or acted upon.

In his book *Expert Political Judgment*, Philip Tetlock presents findings from a large project which he undertook over many years, an investigation of the predictive skill of political experts, based on interviews with 284 experts in the fields of international relations and area studies. His definition of an expert is 'a professional who makes his or her livelihood by commenting or offering advice on political and economic trends of significance to the well-being of particular states, regional clusters of states, or the international system as a whole' (Tetlock 2005:239).

Foxes and hedgehogs

Experts were asked about short- and long-term developments about economic performance, national security, and international events. The findings are sobering. While the experts are performing better than undergraduate students, they fare worse in comparison to a 'mindless' computer algorithm which was supposed to mimic the proverbial dart throwing chimp (a metaphor he introduced and later came to regret after his work had been reduced to this aspect).

However, there is an interesting difference in performance, if only a small one, between the two types of experts. Using Isaiah Berlin's distinction between foxes and hedgehogs (the fox knows many things, the hedgehog knows one big thing), Tetlock found the foxes outperformed the hedgehogs. Although he seems to classify them through self-labelling, he examines their respective

DOI: 10.4324/9781003320227-6

style of reasoning, which turns out to be the most relevant variable in predicting predictive success. While the hedgehog forecasters apply one idea to cover all cases and rely on its explanatory power (when one has a hammer, the whole world looks like a nail), foxes draw on the many little things they know from different cases and various sources of information.

Foxes are better predictors because their reasoning style is more adaptive to rapidly changing events: '[F]oxes doubted that real-world problems could be squeezed, without serious distortion, into syllogistic templates' (Tetlock 2005:140), while hedgehogs want to fit everything into a pet theory or template to analyze every new problem or circumstance.

Foxes seemed to possess a crucial trait that helped them to be resistant to strongly held beliefs: this is the capacity to be self-critical, in technical terms, they are better Bayesian belief updaters. They are more in line with Keynes's famous dictum, When the facts change, I change my mind. In a review symposium on his book, Tetlock replied to his critics in a very conciliatory and insightful manner, echoing the virtues of foxes, trying to learn from his critics. There he acknowledges that not only cognitive styles matter in political debates and advisory skill of experts but also the political biases of the experts and the erosion of trust from political clients and the general public.

Tetlock recognizes that we need to face the problem, 'the degree to which distrust of epistemic communities has become ideologically skewed. Nowadays, even sophisticated conservatives harbor suspicions about the epistemic intentions of behavioural and social scientists whom they see, not altogether incorrectly, as broadly sympathetic to liberal-left causes.' He suggests that restoring 'epistemic trust between conservatism and academia should be a vastly higher priority than it is. There are, of course, no quick fixes, but one promising method of extricating ourselves from otherwise painfully protracted impasses is for respected neutrals to launch a concerted effort to broker what Daniel Kahneman has called "adversarial collaborations"' (Tetlock 2010:471).

In this model, 'rival epistemic and political camps would nominate experts to come together to reach agreements on how they disagree on North Korea or deficit reduction or global warming and then would figure out how to resolve at least a subset of their factual disputes. The disputants would need to specify, ex ante, how much belief change each side would "owe" the other if various agreed-upon empirical tests were to work out one way or the other. When adversarial collaboration works as intended, it shifts the epistemic incentives from favoring cognitive hubris (generating as many reasons why one's own side is right and the other is wrong) and toward modesty (taking seriously the possibility that some of the other side's objections might have some validity)' (Tetlock 2010:471).

Tetlock muses about the incentive structure which leads experts into 'cognitive hubris': 'On one hand, overconfident experts may be more quotable and attract more media attention. On the other, overconfident experts may also be more likely to seek out the attention. The three principals—authoritative-sounding experts, the ratings-conscious media, and the attentive public—may thus be locked in a symbiotic triangle. It is tempting to say they need each other too much to terminate a relationship merely because it is based on an illusion' (Tetlock 2005:63).

Our need for expertise, and our willingness to believe in some kind of expert, is deeply rooted, argues Tetlock: 'We—the consumers of expert pronouncements—are in thrall to experts for the same reasons that our ancestors submitted to shamans and oracles: our uncontrollable need to believe in a controllable world and our flawed understanding of the laws of chance. We lack the willpower and good sense to resist the snake oil products on offer. Who wants to believe that, on the big questions, we could do as well tossing a coin as by consulting accredited experts?' (Tetlock 2005:63).

What kind of music do we hear? It is fair to say that Tetlock does not play the riffs about legitimate or real expertise. On the contrary, he cautions about such ideas and develops a more nuanced tune of the conditions under which expertise operates in society. This includes an awareness of the politicized nature of many expert debates where an appeal to a more forceful assertion of 'real expertise' proves to be futile.

Superforecasters

Tetlock has changed tune somewhat with the publication of *Superforecasters.* Its subtitle announces the shift: *The Art and Science of Prediction.* This book, which he published with his co-author Dan Gardner in 2015, emphasizes the characteristics of those people who show an excellent track record in predicting political events. He emphasizes the style of thinking of superforecasters, which can be summarized in the slogan 'Try, fail, analyze, adjust, try again.' The book is based on research conducted in the context of forecast tournaments funded by the US intelligence community. The tasks set to participants were narrowly defined; they were asked to forecast geopolitical events relevant to the US. The methodology sets out to measure the performance of the forecasts by keeping records of what was said, when. This protocol is borne out of frustration with news pundits such as Tom Friedman, Paul Krugman, or Niall Ferguson, who utter predictions which are couched in wooly terms as to avoid easy falsification.

The research shows that teams of able forecasters ('superforecasters') were much better at predicting events compared to individual able forecasters. These teams were allowed to collect information from the internet or direct communications as they pleased. It appears that one important predictor of success was the lack of formal rules and hierarchies within these groups and an absence of superficial politeness, which could inhibit the frank expression of doubt. In one instance, these groups are called 'anarchist cells.' Despite the better performance of superforecasters compared to 'average' forecasters, the element of luck cannot be eliminated, and towards the end of the book, the authors cite one such successful superforecaster saying, 'I know that luck played a part in the success that I've had' (Tetlock and Gardner 2015:272). Of course, the authors see this as confirmation of the required mindset to do well in such tournaments, in other words, to show humility and grit, and to stick to the eternal cycle of 'try, fail, analyze, adjust, try again.' Sisyphus has to be imagined as a happy man.

The limits of certainty

There are many interesting and funny episodes told in this book. It is also remarkably full with insights about what can realistically be expected of science to contribute to the tasks at hand. Tetlock quotes the mathematician William Byers who wrote 'Most people would identify science with certainty. Certainty, they feel, is a state of affairs with no downside, so the most desirable situation would be one of absolute certainty. Scientific results and theories seem to promise such certainty.' Tetlock comments: 'As the work of accumulating facts proceeds, uncertainty is pushed back. The ultimate goal of science is uncertainty's total eradication. But that is a very nineteenth-century view of science. One of twentieth-century science's great accomplishments has been to show that uncertainty is an ineradicable element of reality. "Uncertainty is real", Byers writes. "It is the dream of total certainty that is an illusion" … Scientific facts that look as solid as rock to one generation of scientists can be crushed to dust beneath the advances of the next' (Tetlock and Gardner 2015:141).

These are important insights and it is good to see our authors endorsing them. But the task at hand is narrowly defined as predicting the likelihood of some future geopolitical event. It has little to do with providing expertise in the sense of advising decision-makers what to do. Instead, towards the end of the book, the role of the forecaster is fused with the role of the decision-maker. This is evident in their critical appraisal of the Prussian field marshal von Moltke. His writings, which were adopted into military manuals of the German army receive much praise, especially for their emphasis on being flexible and using tactics that are appropriate to the situation on the ground. Giving officers enough leeway in decisions of how to achieve a strategic aim allegedly gave the Prussian troops the upper hand in several important battles at the end of the 19th century.

Here Tetlock transfers some of the traits found in superforecasters to perceptive leaders, thereby combining the roles of the expert with the role of the decision-maker. He reports similar anecdotes about Dwight Eisenhower and General Petraeus. But in the end, we are left with little more than platitudes, such as Petraeus's statement that a leader 'needs to figure out what's the right move and then execute it boldly' (cited in Tetlock and Gardner 2015:225).

The book acknowledges the irreducibility of uncertainty when making statements about future events. It draws on the distinction between epistemic and aleatoric uncertainty. Epistemic uncertainty resides in the way we think about reality and the problems of 'getting it right.' It presupposes that reality can be known and we just need to find the best way of getting to this knowledge. Aleatoric uncertainty refers to a reality which is unknowable, and no effort of thinking will get us to the true knowledge. Tetlock uses the metaphors of the clock and the cloud to illustrate the difference: we can understand a machine-like reality like a clock, by disassembling it and by understanding its design principles. But we cannot understand the cloud in the same way as it has an inherent unpredictability. An accurate weather forecast of ten years, ten months, or even ten weeks is therefore impossible.

Curiously, the book does not make the most of this insight. In the many examples discussed, we are led to believe that reality is cloud-like, not clock-like. We are, therefore, in the realm of aleatoric uncertainty. Nevertheless, we are also given the impression that superforecasters are better at predicting future events than everyone else, which suggests that either reality is more clock-like or that some people can get better insights into reality via epistemic achievements. We have seen which kind of mental condition is required ('try, fail, analyze, adjust, try again') and the book summarizes the traits of superforecasters in the appendix, appropriately entitled Ten Commandments for Aspiring Superforecasters. Here we are given another take on the clock and cloud conundrum. The first commandment says 'Focus on questions where your hard work is likely to pay off. Don't waste time either on easy "clocklike" questions (where simple rules of thumb can get you close to the right answer) or on impenetrable "cloud-like" questions ... Concentrate on questions in the Goldilocks zone of difficulty, where effort pays off the most' (Tetlock and Gardner 2015:277). The aspiring superforecaster is given some clever marketing advice to position herself in such a way that reasonable expectations are met, illustrating my distinction between well-defined and ill-defined problems.

In large parts, the book forgets about the uncertainty problem altogether. It can be characterized as one huge undertaking in developing metrics that quantify probabilities, something which runs counter to the logic of uncertainty. The superforecasters that receive the most praise and coverage in the book are those that would specify their attribution of likelihoods of certain events in a very precise way: not 60/40 but 65/35, or even more granular. The authors are aware of the potential damning criticism, made by Taleb and others: 'what matters can't be forecast and what can be forecast doesn't matter' (Tetlock and Gardner 2015:238).

A sober view of experts

The psychologist Mark Burgman has written another perceptive account of expert knowledge and expertise, covering similar ground as Tetlock. However, he is not so much interested in the question of what makes a good forecaster but 'how a decision-maker should learn from experts' (Burgman 2016:3).

At the outset, Burgman leaves no doubt about his critical approach to the high esteem experts are hold in and the high esteem they confer upon themselves: 'Society generally accepts that scientific and technical experts provide a unique, valuable resource. The US National Research Council, for instance, asserted that scientific experts have indispensible knowledge, methodological skills and experience. And scientists themselves believe it. For example, a review of expert veterinary epidemiologists stated "[e]xperts can be excellent reservoirs, integrators and interpreters of knowledge. In many settings, their ability to generate accurate predictions is a critical function of their profession and a key measure of their success: for example a stock broker's ability to forecast performance of a market, or a physician's ability to triage and assess a patient's need for hospitalization"'(Burgman 2016:4–5).

Burgman does not buy into this hype. He promises the reader to 'examine this general claim as well as the specific performances of stockbrokers and physicians. In many circumstances, such optimism is misplaced' (Burgman 2016:4–5).

He intones the tune of problematizing official expertise by asking pertinent questions about its role of authority in society.

He provides evidence for the overinflated reputation of technical experts. They regard their ability higher than is borne out by their performance in standardized tests, and they are regarded by laypeople to be far superior to themselves, even when there is no gap in performance. Not recognizing the limits of one's knowledge makes experts over-confident: '[I]In general, experts lack humility about what they do not know, which can make them dangerous' (Burgman 2016:70). He draws attention to the fact many white males belong in this category.

There is a problem with experts then. Experts assume a position of authority which can intimidate people. And society may, deliberately or unwittingly, foster this position because we tend to assume that there is a right answer to any given problem and that experts are the best people to find it. In Burgman's words, 'we just need to consult the right oracle to discover it'. However, when we are facing uncertainty different expert opinions will come to the fore and expertise gets easily mixed up with advocacy. In addition, the high social prestige experts are enjoying also means that society tends to exclude lay persons' opinions on controversial matters, even if they have relevant technical knowledge. 'What counts as expertise depends on context. If experts are tested then expertise from all domains may be considered, including what may be considered lay knowledge. Still, it is commonplace for those regarded as "non-experts" to be left out of the conversation, often to the detriment of the decision' (Burgman 2016:22).

We can clearly hear the tune of deconstructing exaggerated epistemic claims of expert authority. This does not mean that official experts or official expertise should be discarded. The argument is one about inclusion of potentially useful voices which are normally marginalized.

The high social status of experts is also challenged via statistical methods which have been used in computer-assisted programmes. Paul Meehl's provocative book *Clinical versus statistical prediction: a theoretical analysis and a review of the evidence*, published in 1954, showed that clinical diagnoses by doctors and decisions made by parole boards were less accurate compared to algorithms. Burgman comments: 'Despite decades of consistent research findings in favour of the statistical method, most professionals continue to use subjective, clinical judgements and so not use quantitative tools, even when they are available' (Burgman 2016:17).

The tune gets complicated, it receives a twist by introducing algorithms. Burgman thus not only allows for the presence of laypeople in deliberations, but he also welcomes the contribution made by AI. However, this is a contested terrain, see Dressel and Farid (2018) and Lin et al. (2020). I will return to this issue in due course.

Forming judgements in the face of uncertainty

Like Tetlock, Burgman distinguishes between different types of uncertainty. He differentiates between three types: natural variation, knowledge-based uncertainty, and language-based uncertainty. Like Tetlock, he moves on to probabilities in order to quantify what essentially are chance events. Probabilities express the level of subjective confidence of experts in their opinion on a matter which involves uncertainty. Collecting more data helps to reduce uncertainty only where a lack of knowledge is at stake. Natural variation cannot be reduced by collecting more data. This means that we cannot do without subjective judgements of experts when there is a need to act but insufficient data to evaluate the consequences of the action (Burgman 2016:34).

We encounter linguistic uncertainty in a famous example from meteorology. When a weather forecast states that there is a 30% chance of rain, this could mean at least one of three things: (1) It will rain tomorrow in 30% of the region; (2) It will rain tomorrow for 30% of the time; or (3) It will rain on 30% of the days like tomorrow (Gigerenzer et al. 2005). While weather forecasters normally assume the latter, this is rarely stated and the public does not necessarily see it this way or understand the reasoning behind it.

Of the three types of uncertainty, only the natural variation type ('aleatoric,' in Tetlock's parlance) poses a serious problem. Epistemic uncertainty can be addressed through more and better data and models and language-based uncertainty through better science communication. Natural variability, like Kant's unknown thing-in-itself, is the hard case. There are limits to what can be known about many things between Earth and the sky. However, psychological research on expertise, as the contributions of Burgman and Tetlock show, tries to move on from the hard case to the instances where we are dealing with (subjectively assigned) probabilities. This leads the inquiry to examine the question of how good expertise is in specific cases and what makes a good expert, which means finding the personal traits which make someone good at predicting future outcomes.

Burgman takes a different approach to Tetlock when pondering the best way of setting up expert groups who make predictions. Whereas Tetlock put together superforecasters who then deliberate, ideally in a spirit of friendly criticism, Burgman focuses on the social composition of the group. He warns against simple aggregation and unstructured deliberations as these will lead to groupthink. 'If we acknowledge that expert judgements typically are better than non-expert ones and that the notion of expertise should be broadened to include people who do not fit the conventional mould but who may have valuable knowledge and experience, then perspectives from a broad cross section of people with relevant exposure to the problems may provide a relatively reliable estimate or prediction' (Burgman 2016:131).

Groups of experts should be stratified according to social context, motivation, technical background, gender, and culture. Burgman reckons that in many cases, the average or median of a group estimate provides a reliable summary of the expert opinion. However, this does not mean it is right. Differences of opinion should, therefore, not be discarded but fed into the discussion—maybe those who assume extreme values have good reasons for doing so or know something that other group members don't.

This chapter has given a short overview of some important insights from authors that have examined the claim that professional and scientific experts can make better predictions than ordinary people. This claim does not stand in its general form, as both authors have shown in different ways. I will return to the issue of forecasts in part 2 of this book. Forecasts rely on assumptions, theories, and models. Models are sometimes implicit, like mental models, sometimes based on codes and algorithms and running on computers. Climate change, Covid-19, and the human and technical attempts at forecasting in professional contexts are prime examples.

In the next chapter, I will discuss the other side of the relation between expert and decision-maker and focus on the role decision-makers play vis-à-vis expert advice.

Chapter 5

The Politics of Knowledge

Beyond the linear model

The relation between knowledge and politics has been the subject of many publications over the centuries, scholarly and non-scholarly alike. The term knowledge politics, or politics of knowledge, is useful if ambiguous. It denotes the stimulation and regulation of knowledge production, as, for example, in science policy. It also denotes the active role of knowledge, as expressed in Francis Bacon's famous thesis that *scientia est potentia*, which frequently, but questionably, is translated as *knowledge is power*. Two influential thinkers of the 20th century, Max Weber and Michel Foucault, have emphasized and examined the close link between knowledge and power, and both have shaped the dominant view that knowledge *is* power. While Weber develops this idea through his focus on the technical expertise assembled in state bureaucracies (Weber 1978), Foucault asserts that power and knowledge are inextricably bound up with each other so that no power can be exercised without knowledge. Conversely, there can hardly be knowledge without power implications (Foucault 1980).

In public policy debates, a version of this dominant model has emerged, which is called the linear model of policymaking. In this model, there is a linear connection between the creation of new knowledge and its ultimate application in policy. Sometimes this linear view is wedded to the assumption that better decisions emerge as a result of better knowledge. This has been called the 'rational' view of policymaking.

Others have argued that such views overstate the role of knowledge in politics (Adolf and Stehr 2017). Social scientists have submitted several of its underlying assumptions to critical scrutiny. Critics have argued, for example, that policies are not routinely based on the best available knowledge or evidence or that bodies of scientific knowledge do not necessarily lead to courses of political action, let alone to the 'best' policy (Jasanoff 1990). Scientific knowledge is often irrelevant to policymaking, and policies are based on cherry-picked knowledge claims for political reasons. This is to say that knowledge has a legitimatory function (Habermas 1971).

DOI: 10.4324/9781003320227-7

Social scientists are divided on the question of what influence knowledge has on politics. On the one hand are the supporters of a rational policy concept, according to which more and better knowledge leads to improved policy outcomes. This is often accompanied by the assumption that an epistemic consensus would make political action easier (Cook et al. 2013; Haas 1992). On the other hand, it is argued that policymaking is largely divorced from scientific evidence but that this need not be a disadvantage (Collingridge and Reeve 1986; Pearce et al. 2017; Sarewitz 2000).

Knowledge, science, and expertise

Returning to the definitional point from above, it seems important to spell out what one means by the terms knowledge, science, and expertise. Not all knowledge is scientific knowledge: there is traditional knowledge, practical knowledge, tacit knowledge, and so on (Grundmann and Stehr 2005). To bring scientific knowledge to bear on the policy process, several transformations have to take place, assuming scientific knowledge is the sum of published research. Decision-makers judge for themselves which knowledge they will use as a basis for their decisions: political power always decides how to use expertise for legitimation. Science, as the sum total of published research, cannot 'speak' to politics (or anyone else), only scientists, and groups of scientists can do so. This means scientists make judgements about the existing literature and therefore give their expert opinion (Turner 2003).

A logical and practical difference exists between the sum of published research and the opinion of scientists/experts, no matter how well they are trying to do justice to the published literature. The politics of knowledge denotes the more or less skilful exploitation of this fuzzy zone between the two. At times, the scientists/experts who are called upon do not have qualifications in the field they are assessing for policy purposes. In these cases, their credentials make up for this shortcoming, and decisions are presented as being based on 'science' because highly reputable scientists sit on an expert committee or have lent their name to a course of action. The perception and attribution of cognitive authority is a relevant social mechanism to solve the impasse caused by scientific debate and disagreement.

If one adopts a wider definition of knowledge, which is not restricted to scientific research, the notion of the expert and of expertise becomes relevant. The function of experts is to translate existing knowledge and to align it with decision contexts so that specific courses of action can be designed (Grundmann 2017). This is much broader than the science-based view, and it emphasizes the relational and pragmatic nature of expertise. No matter which of the two views one prefers, an important asymmetry exists between knowledge and politics. While knowledge may influence the ways we think about the world, its problems, and possible solutions, it is politics that decides which parts of this knowledge, with what implications, are selected for political action. In other words: scientific experts may be able to select problems for attention, and frame problems in a specific way, thereby playing a quite influential role. But they usually do not have the political power to take decisions.[1]

There is another possibility for changing reality through science. Different social actors may develop new applications of knowledge through technological innovation. This is the norm in democratic societies, which are based on the freedom of enterprise. Unless something is explicitly forbidden, it is allowed within existing regulations. Hence private companies may introduce innovative products that later may turn out to have unforeseen, even negative impacts or at least the potential for such impacts.

De Vries (2007) tells the story of Dutch researchers who piloted blood tests, offering pregnant women to participate in trials to find out about potential problems with their unborn child. They created a fait accompli, circumventing existing legislation. De Vries, following Ulrich Beck, calls these events instances of 'sub-politics.' A new reality is created via the private sector. No one has planned ubiquitous computing or the car dependency of modern societies. These are the results of self-organizing co-evolutionary processes in which private initiative successfully embarked on new business opportunities which were not resisted politically but often politically welcome and even supported by the government (Mazzucato 2011). In this sense, expertise in the service of private companies does have societal and political impact. Where it raises controversies, experts employed by companies, universities, and state administration become embroiled in debates about the risk which is posed to us as a result.

The special role of pharmaceutical companies during the Covid-19 pandemic is another prominent example. Despite efforts at coordinating responses to the spread of the disease via the WHO and COVAX, private companies have produced and marketed vaccines that have made a difference. They have made a difference in that the effects of the pandemic have been softened, even if the expectation to eradicate the virus has been disappointed. But they have also made a difference as these companies have changed the social and political reality in that decisions by public institutions have been pre-empted by private initiative.

Political science has made several contributions to the problem of expertise. Here I will focus on the concepts of epistemic communities, deliberative democracy, and knowledge regimes.

Epistemic communities

Peter Haas has put forward a much-acclaimed model to explain international policy coordination. He notes that 'speaking truth to power' is a fraught exercise: 'Even when scientists think they have developed truths for power, power appears disinterested at best, and possibly even uninterested' (Haas 2004:570). So under which conditions does power listen to truth? It is a rare phenomenon, Haas tells us (but not so rare as to be entirely dismissed, as is practice among many policy analysts). One precondition is that scientific knowledge is transformed into 'usable knowledge,' which is 'accurate information that is of use to politicians and policy makers.' Echoing the linear model, Haas postulates 'that science must be developed authoritatively, and then delivered by responsible carriers to politicians.' This is done via a transmission belt 'of like-minded scientists … called an epistemic community' (Haas 2004:576).

He defines an epistemic community as a 'knowledge-based network of specialists who share beliefs in cause-and-effect relations, validity tests, and underlying principled values and pursue common policy goals. Their orientation is perhaps best expressed in the words of one member, who voiced his willingness to accept the "plausibility of a causal link without certainty"' (Haas 1992:187f.). In another place, he writes: 'Because of their environmental values, the members of the group advocated anticipatory action despite the range of uncertainties' (Haas 1993:176).

Haas assumes that 'the more autonomous and independent science is from policy the greater its potential influence.' First, science needs to be developed behind a politically insulated wall and then transmitted by epistemic communities to decision-makers. He further assumes that decision-makers recognize 'the limits of their abilities to master new issues and the need to defer or delegate to authoritative actors with a reputation for expertise' (Haas 2004:576). It has been commented that 'Haas is encouraged by the increasing number of scientific staff in government' and the 'increasing deference paid to technical expertise and in particular that of scientists.' However, at the same time, it has to be noted that 'data are usually ambiguous and scientists are usually divided over public issues, concerning, for example, the environment and energy' (Jacobsen 1995:302–303).

When are epistemic communities influential? Adler and Haas (1992:380–381) argue that 'where policymakers have no strong preconceived views, and the issue is a first foray, then these bands of experts will exercize a lot of influence.' However, as Jacobsen rightly points out, 'if the issue does not matter very much, then experts do – not an appealing analytical agenda' (1995:303). In defence of Haas and Adler, it must be said that experts in such situations can play an agenda-setting role. Their ideas are instrumental in getting an issue onto the political agenda. This point is conceded by Jacobsen when he states that '[b]y persuading players to accept a 'definition of the situation,' one gains some control over the outcomes because how a problem is defined determines the nature of the solution' (Jacobsen 1995:292). 'The greater the uncertainty, the more influential is the role of ideas' (Jacobsen 1995:293).

There are two basic assumptions in Haas's framework: that policymakers turn to experts to reduce uncertainties and that policymakers act on the basis of expert consensus.

While the first assumption may be valid, the second does not follow from the first. Often there is more than one expert opinion, and the decision-maker may wish to ignore the advice given. The decision-maker may even manipulate the process of expert elicitation in order to get the desired advice, as we have seen above in the section on Knowledge Politics.

Another problem resides in the very idea of an epistemic community. There is a danger that the community reinforces its beliefs even when there is evidence contradicting these beliefs. The epistemic community can easily fall prey to group think. If this is the case, critical thinking and search for different explanations, and alternative courses of action would be suppressed. As a result, we should be careful about the advice produced by epistemic communities.

There is an interesting link to the discussion about superforecasters in the previous chapter, especially Burgman's call for a socially diverse group of experts. At first glance, although Haas appears to join those who are riffing about expert authority and legitimacy, his views are more nuanced than that. Most importantly, he thematizes the importance of values when experts make up their opinion.

Deliberative democracy

A large body of literature, influenced by Habermas's work on the public sphere and the normative ideal of the 'ideal speech situation,' including the 'power of the best argument,' has suggested ways to reconceptualize the role of public participation in political decision-making. These approaches can be summarized under the label of deliberative democracy. The work of John Dryzek is probably the most salient in this literature (Dryzek 2002; see also Fischer 2003; Krapohl 2003; Mansbridge et al. 2012; Renn 1999; and Thompson 2008).

Pellizzoni (2001:66) summarizes the main features of the deliberative model in what he calls 'three virtues.' The first is the civic virtue. It is assumed that discussion produces 'better' citizens: 'individuals who are more informed, active, responsible, open to the arguments of others, co-operative, fair, able to deal with problems, ready to alter their opinions.' The second produces legitimation for the political system: 'a decision taken following open discussion has greater legitimacy. It is more likely to be respected because it has been freely endorsed.' The third is related to the open discussion of arguments, with a view to restrict strategic behaviour.

Pellizzoni interrogates the deliberative model, reaching the conclusion that the argument about the power of the best argument is a myth. However, this does not mean that all arguments are equally valid. Habermas makes too strong assumptions about the logic of reaching an agreement, inter alia that 'there is only one "right" response to a problem, and that the parties are able to communicate in order to reach agreement upon it' (Pellizzoni 2001:71).

In particular, Pellizzoni sees some deep level assumptions of Habermas as problematic:

> For Habermas, this structure is evidenced by the fact that concepts like truth, rationality, foundation, consensus, perform the same grammatical

> roles in all language communities so that there is semantic identity between words and propositions. The linguistic presuppositions of discourse permit the passage, in a controversy, from the pragmatic to the ethical level, and from the latter to the moral one. The incommensurability of the terms of a conflict can be overcome if those concerned shift from the values relative to the good life to those of justice. At the moral level, reason can be brought to unity. But is this 'deep' level truly important?
>
> (Pellizzoni 2001:71)

Pellizzoni denies the importance of these assumptions. Instead, he argues that the 'presence of irreconcilable alternatives is a crucial problem for deliberative democracy.' Incommensurable views cannot be overcome by the logic of the best argument. Participants in such controversies are typically not able to agree on just procedures to resolve the issues. But this does not thwart the search for a deliberative solution. According to Pellizzoni, such a solution has to be pragmatic, not based on the strong assumptions of Habermas: 'A willingness to act jointly does not signify a willingness to co-operate "dialogically", a readiness to consider the reasons of the others and to change one's own convictions. It is not necessary for the participants' viewpoints to coincide for co-operation to come about: all that is required is that their individual goals dovetail together. This may be the result of strategy' (Pellizzoni 2001:77–78).

In other words, it is futile to aim at excluding strategic behaviour from the acceptable repertoire of social action when confronting situations of deep value conflicts and uncertainty. And it is futile to expect alignment of beliefs, reasons or world-views in order to agree on a common course of action.

Fuller goes further in his critique of deliberative democracy theorists, 'who seem to want to ensure that however many people, of whatever background, are involved in democratic processes, they should end up taking decisions that are "correct" by some independently approved normative standard, typically defined as closeness to one's own true interests—if not closeness to some more comprehensive understanding of "objective" truth, for which fidelity to the "facts" serves as a proxy. Yet, such lofty metaphysics masks a technocratic "rule by experts" …' (Fuller 2021:352).

In a similar vein, Lövbrand et al. (2010:485) contend that for deliberative democracy theorists 'only one true response to a problem exists. While there may be many private reasons, there is only one public reason … The better argument is consequently the one that analyses a problem most thoroughly and indicates the optimal solution in terms of technical excellence and moral rightness.'

They point to the radically different take from some authors in STS which have a deep mistrust in universal solutions. 'By studying how claims to epistemic validity come into being and are sustained across cultural and political contexts, many scholars of science and technology have questioned the notion that all publics reason in the same fashion or from the same epistemological foundations.' As we shall see, these differences across political cultures have to be taken seriously. More generally, the STS critique aligns with views of politics as an agonistic field (Connolly 2002; Mouffe 2005), and rejects the ideal of reaching a consensus as the most desirable goal: 'Lacking a shared understanding of the meanings of principles and concepts, diverse publics cannot be expected to reach agreement on what is good and true' (Lövbrand et al. 2010:485).

Berg and Lidskog (2018:13) praise the virtue of deliberative democracy without endorsing the strong assumptions made by Habermas. They apply the principles of this approach to global politics, where they are supposed to solve the lack of legitimacy, given the 'lack of democratic institutions and sovereignty at the international level.' Mutual understanding between different actors at different levels of the international system is key. Here the authors come very close to the approach taken by Burgman. The need for dialogue and cooperation among members of diverse teams is what links them together. As Mansbridge et al. (2012:12) put it, 'The central democratizing aspect of deliberation is the inclusion of multiple perspectives in a communicative and equitable manner.'

Berg and Lidskog (2018:13) comment that 'essential components of deliberative theory are thus inclusion and communicative equality and authenticity. Furthermore, it is crucial that deliberation aims for some type of shared agreement about how to move forward in order to enable decision-making.'

Berg and Lidskog use the examples of specific settings in international environmental politics. The challenge is to find, and to apply procedures of inclusion, also with regard to scientific assessments. Inclusion is desirable because it is functionally superior to other processes of decision-making.

Using the example of the Intergovernmental Panel on Climate Change (IPCC), the authors indicate that the 'inclusion of researchers with different backgrounds may thus be encouraged because it is favorable to a multiplicity of perspectives.' However, they warn to be aware of the uneven representation of perspectives and of what is lost in the process of translating and representing perspectives. They cite the work of deliberative democracy theorist Dennis Thompson: 'From a deliberative systems perspective, equal participation requires that no perspective, group, institution, or sphere completely dominates the process through which reasoning is presented' (Thompson 2008:504–505).

As indicated, Berg and Lidskog emphasize the need for a multiplicity of perspectives: 'what is crucial for deliberative capacity is not group- or interest-based representation but the inclusion of a multiplicity of perspectives, thus promoting quality deliberation and robust policy solutions' (Berg and Lidskog 2018:13).

Nevertheless, two problems remain. The first is that some perspectives will not be represented, and some people do not want to be represented, as Iris Young has convincingly argued (Young 2001). This is bad news for deliberative democracy theorists as it means to either fall back on an aggregative model of representative democracy to which the deliberative model was advertised as an alternative (Lövbrand et al. (2010). Or it means to exclude those people, interests and perspectives because they fail some standards of moral conduct (Grundmann and Mantziaris 1991).

The second problem is that we are never on a level playing field. There is always someone who defines the rules of the game and thus has an influence about who is included or excluded and which perspectives are foregrounded or backgrounded. As Young (2001:677) observes, 'elites exert their power through managing deliberative settings.'

Deliberative democracy is always situated in a pre-existing institutional framework. This means that the rules of the game and the roles specific actors can play are, at least to some extent, pre-defined. This means powerful actors decide whose opinion counts including new representations, new actors, and new voices. Steve Fuller (2018:139) uses the term modal power to describe this setting: 'Modal power consists in the capacity to decide what is and is not possible,' reminiscent of Schattschneider's insight about the role of controlling alternatives in politics.

The above seems to suggest two things. First, deliberative democracy argues for the inclusion of lay participants in decision-making because this will lead to a better society, as it entails better decisions that are supported by informed citizens. In so doing, it questions the inherent superiority of professionals, specialists, and other 'official' experts. Citizens should have their say, and the merit of their arguments should be taken into account and taken seriously. This normative model has influenced a large follower base, even if the more demanding elements of discourse ethics have not been adopted by some proponents of deliberative democracy or by other authors discussed above. The very idea has proven attractive. It appears in many different approaches that want to go beyond the status quo and pose the question of what is, and what is not possible.[2]

There is a lingering doubt referring back to the critics' point of deliberative democracy. The litmus test of truly democratic procedures would be to give up on ideas about 'correct' outcomes that ought to be preferred. One cannot have it both ways: either substantive policy goals are fought for, or procedural steps are followed wherever they may lead.

Looking back at this section in musical terms, we can hear another variation on the theme 'experts are not enough to reach good decisions.' The theme is in the same key as the contributions from the literature on lay expertise, Post-Normal Science, and the diversity of expert groups.

Knowledge regimes

Compared to the previous section, which involved issues bordering on moral philosophy, here I am presenting work that is located at the level of policy analysis. Campbell and Pederson have suggested a framework for comparative, empirical studies about policy-making where expert advice is crucial. They use the term knowledge regimes to describe such settings and define them as 'sets of actors, organizations, and institutions that produce and disseminate policy ideas that affect how policy-making and production regimes are organized and operate in the first place. Knowledge regimes are important because they contribute data, research, theories, policy recommendations, and other ideas that influence public policy ...' (Campbell and Pedersen 2010:167).

Campbell and Pedersen (2015) use the example of economic policymaking since the early 1980s in the USA, Denmark, France, and Germany. Here they demonstrate that each knowledge regime is nationally specific. As they explain, 'Knowledge regimes are the organizational and institutional machinery that generates data, research, policy recommendations, and other ideas that influence public debate and policymaking.'

Knowledge regimes are made up of 'policy research organizations like think tanks, government research units, political party foundations, and others that produce and disseminate policy ideas.' This reveals an important limitation of such regimes as they comprise only parts of the academic and political elite of a country. Citizen expertise is not part of such arrangements.

Participants of such knowledge regimes 'engage in similar activities and interact more frequently with one another than with organizations outside the field' (Campbell and Pedersen 2015:3).

Still, the conceptual framework of knowledge regimes allows us to adopt a broader perspective when analyzing actors in the policy process compared to much current literature which adopts a rather narrow focus on science, scientific knowledge, and scientific expertise.

But back to the national institutional context in which knowledge is produced and applied, in other words, how expertise is mobilized in politics. There are interesting differences across countries when it comes to the role such knowledge regimes play in society. For example, Germany is a country where state administration is highly dependent on in-house expertise of which a large proportion is based on legal expertise. Members of Parliament tend to have a law degree (Kintz 2010; Speth 2016), and most legal initiatives are written with legal-procedural matters in mind. In contrast, in the UK only 14% of MPs have a law degree, whereas 37% come from business and finance, and another 18% from media and PR (Hackett and Hunter 2015). These socio-demographic facts have undoubtedly an effect on the form and possibly the content of law and state regulations.

The USA has a vast number of think tanks, some of which are independent, some combine lobbying with expertise, and some are open advocacy outfits (Stone and Denham 2004). The UK and US have a much higher number of think tanks compared to Germany, where the most important think tanks are aligned with political parties. German think tanks are more dispersed across the country compared to the centralized structure in the UK and US, where most think tanks are close to the centre of political power in London and Washington, DC (Thunert 2004).

An important aspect is the above-mentioned issue of in-house expertise: some governments (I mentioned Germany) are keen to create their own knowledge base on virtually all important political issues. This is part and parcel of public administration, supposed to help counter the information asymmetry that appears when private actors or science organizations make claims that would, in effect, restrict political room for manoeuvre. The civil service in the UK can be a powerful bastion in advocating or resisting policy change, as Peter Hall has shown in his analysis of the role of the Treasury (Hall 1993).

In a different context, Collingridge and Reeve touched on this very point. They emphasize that decision-makers are used to decisions under uncertainty, they do not try to collect comprehensive data before making a decision (see also Lindblom (1959). In this situation, it is advantageous to reference scientific knowledge claims, not least because all lobby groups in a policy arena tend to use them and because scientific knowledge has higher prestige than other forms of knowledge.

> The role of scientific research and analysis is therefore not the heroic one of providing truths by which policy may be guided, but the ironic one of preventing policy being formulated around some technical conclusions. Research on one hypothesis ought to cancel out research on others, enabling policy to be made which is insensitive to all scientific conjectures.
>
> (Collingridge and Reeve 1986:32)[3]

However, their distinction between two modes of decision-making (under- and over-critical) seems too rigid to cover the dynamics of the politics of knowledge. After all, sometimes, we do get policy decisions after a period of seemingly endless technical debate. And there are examples where a policy consensus is undermined by emerging knowledge claims, as witnessed by constant scientific and technological innovation. Nevertheless, the authors make an important point that in modern societies, knowledge is likely to be used in legitimizing or blocking specific political decisions. *Knowledge does not prescribe policy, but either legitimizes or prevents it.*

The framework of knowledge regimes could shed a new light on the case studies of climate change and Covid-19, which will be discussed later. Here is why I think it could be important: Campbell and Pederson say that many scholars have recognized the role of ideas in politics (following Max Weber's insight), but they do not study how ideas are produced and disseminated and ultimately become part of public policy.

They say: ' [T]his is an age when policymakers strive to recognize and improve their country's institutional competitive strengths and rely increasingly on the production of policy-elevant knowledge to do so.' The historical period they are talking about is the end of the so-called Golden Age (post-war period) and the onset of Globalization. Climate change and Covid-19, in different but similar ways, exemplify the challenge to governments of replying to a problem based on their institutional setting and resources to improve their competitive position.

Campbell and Pedersen argue that the spread of neoliberal economic policy ideas was uneven across countries, contrary to the view that there is a uniform neoliberal hegemony. In a similar way, one could say that governments have pursued very different approaches with regard to climate change and Covid-19, both eminent global problems. They have not reformed their knowledge regimes in response to a crisis—and by this, I mean the failure to respond to the problems in a timely and effective way. The theory of knowledge regimes would predict that these regimes change over time when challenged by significant developments. Globalization was the challenge studied by the authors, the question is if similar developments are to be expected with climate change and Covid. As noted, the lack of a wider representation of societal voices needs to be kept in mind.

Notes

1 The example of the German health expert Karl Lauterbach is exceptional. He combines the roles of scientist, expert, and politician. He holds various academic degrees and is a registered general practitioner. As a health expert for the SPD, he made public comments, an activity that became pronounced during the Covid pandemic. He was present on most talk shows in German TV, sometimes several days a week, warning of the dangers and advocating strict measures. He gained celebrity status, which helped him landing a cabinet post. When Olaf Scholz formed his government in December 2021, Lauterbach was made health secretary, the only member of government outside Scholz's inner circle.

2 This would require a much more in-depth discussion which I cannot provide here. See Gunderson (2018) for an account of important aspects.

3 This is similar to Max Weber's statement that 'the ruler … keeps one expert in check by others' (Weber 1978:995).

Chapter 6

Expertise and Economics

A critique of expertise

This chapter examines Roger Koppl's book *Expert Failure* which contains an analysis of experts from am economic perspective. Koppl locates the problem of experts in the context of a division of knowledge in a society where an asymmetric distribution of knowledge can lead to an (unwarranted) dominion of an expert elite over other groups in society. This asymmetry emerges when experts not only give advice but have the power to prescribe actions for others. For Koppl, this is wrong in principle as he sees it as a form of exercise of power in society. Professions are usually licensed by state authorities and therefore represent government intervention ('regulation') into markets. Koppl is convinced that such interventions, often made with the best intentions, provide worse outcomes compared to markets which have been set up properly.

This framing of the problem leads him immediately into normative terrain and he reckons that one may take a largely favourable or sceptical view of experts (Koppl 2018:9). It is clear from the outset that he is critical of experts. This is based on a critique of views going back to the Socratic tradition, which saw experts as 'being epistemically and morally superior.' Here Koppl is in line with arguments presented above. He says, '... knowledge is not hierarchical, unitary, explicit and bookish. It is, instead, generally emergent from practice, often tacit, and embodied in our norms, habits, practices, and traditions' (Koppl 2018:11).

He also shares with some Science and Technology Studies (STS) scholars the view that experts are neither more nor less honest or selfish than non-experts (the position of Harry Collins is notably different in making special claims for scientists). Where he goes beyond what one could call the STS platform is when he explicitly defends a libertarian position which sees markets as the solution to many problems arising from the division of knowledge and, most importantly, a potential failure of experts. He pithily states, 'Expert failure is more likely when experts choose for their clients than when the clients choose for themselves' (Koppl 2018:12).

The solution to this problem is the design of market structures which generate more reliable expert opinions. This will be achieved, according to Koppl,

DOI: 10.4324/9781003320227-8

through an institutional setup in which experts compete with each other and in which the clients are free to choose which opinion to believe. This is the true meaning of pluralist democracy and the rule of law, both concepts which are incompatible with the rule of experts, according to Koppl. Note, again, the emphasis on the diversity of expert groups.

Musically speaking, Koppl presents a rather simple riff, not a tune. This may be regarded as a counterpoint to Oreskes, given its simplicity and radical posture. The riff does also come with a coda, although one which complicates the original tune.

He defines an expert as 'anyone paid for their opinion'—which seems brief and elegant. It resonates with my relational definition of experts who are providing advice to a client. The definition presupposes a setting of supply and demand of experts and of expert advice. But this need not mean that the relation is primarily a monetary relation. Advice can be given in exchange for non-monetary items; it also can be provided 'for free.' Koppl's value focused definition, while not surprising for an economist, seems flawed. It is also unnecessary as it obscures the important social relation at stake. The important element of the social relation is not about payment but about advice. From the client's perspective, the most important question is if the expert can be trusted. Paying an expert does not increase the quality of the advice or its reliability—it merely creates a pre-commitment on the side of the client (Sniezek et al. 2004).

Koppl's definition of experts as 'anyone paid for their opinion' raises at least two questions: What about experts who do not get payment? Are they, *per definitonem,* not experts? And what about experts who do not offer an opinion but execute a job which requires high degrees of knowledge and skill and little latitude for deviation from routines?

An example of the first question is the IPCC and of the second professional organizations. As I discuss these examples in greater detail in Chapters 3 and 7, here I restrict myself to some brief points as they emerge from Koppl's treatment.

Members of the IPCC do not receive remuneration apart from travel costs and per diem payments. In this sense, it is not a body whose members get paid for their opinion. Koppl mentions the IPCC in passing without applying his theoretical apparatus. Instead, he introduces the ad-hoc term 'big player' to describe the IPCC. This term was previously used by Butos and McQuade (2015) to analyze the IPCC, and the term was coined by Koppl and Butos in 1993.

Here is the definition of a big player: 'Big Players, as described by Roger Koppl in an economic context, are "privileged actors who disrupt markets" in the sense that although they are not subject to market constraints or to the discipline of market competition, their discretionary actions have widespread impacts on market participants' expectations and actions … In markets, prototypical Big Players are central banks and government agencies empowered with discretionary policymaking. As Koppl argues, markets dominated by Big

Players are prone to herding, where market participants, with little reliable information about the Big Players' next move, look to what others are thinking and doing' (Butos and McQuade 2015:168–169).

This may provide some insights about the central role of the IPCC in public discourse and it can be read as akin to the sociological theory of bandwagon effects (Granovetter 1978). However, we get little new insights about the specifics of expertise offered by different types of expertise. While the observation about the IPCC is not mistaken, there is little on offer in terms of conceptual work about different types of expertise. However, we do get some strong normative comments regarding the role of professional experts.

As I have noted earlier, professional organizations are providing expertise, and its members often make decisions for others. I call this function of expertise 'specialist.' For Koppl, this is wrong in principle as he sees it as a form of exercise of power in society. Professions are usually licensed by state authorities and therefore represent government intervention ('regulation') into markets. Koppl is convinced that such interventions, often made with the best intentions, provide worse outcomes compared to markets which have been set up in the proper way. There is little evidence provided to support such a sweeping statement. To be fair, he acknowledges that markets do not work well outside rules that provide incentives.

Methodological and political individualism

In Koppl's view, the ideal situation is one where the citizen is offered the appropriate information in order to make a decision for herself. Consumer tests and the like are the institutional form which delivers the necessary information. In my view, this leads to a very problematic outcome. We end up in a situation where the state, and its professional organizations, have abdicated any responsibility for the safety and security of its citizens. Imagine the issue of food safety. Without standards and licenses, I would spend hours in supermarkets equipped with a magnifying glass and an internet search engine to establish the ingredients on the packaging. I would need to find out if the specific food item contains some stuff that could have adverse impacts on my health. This presupposes I would need be able to tell which of the ingredients has a potentially harmful effect on my well-being. To be sure, we already find ourselves in such situations, especially if we suffer from food allergies or have ethical reservations about certain properties of the item. Koppl wants to make this the norm, all in the way of 'empowering the layperson.'

Public health issues are another example where the limits of individual liberty are obvious. As the case of Covid-19 exemplifies, the logic of pandemic diseases requires adherence to specific rules and restrictions in order to keep them at bay.

The limits of Koppl's economic analysis should be plain. A sociological perspective, by contrast, sees different forms of social coordination, all with their advantages and disadvantages: markets, hierarchies, and voluntary organizations. Even economists have discovered the benefits of hierarchies (Coase, Williamson, etc.), but Koppl does not engage with this radically different intellectual agenda.

The title of the book indicates that expert failure is its central theme. Unfortunately, there is no clear definition of the term. Expert failure seems to be used synonymously with expert error, thus highlighting the potential epistemic shortcomings of experts. However, Koppl says that his interest is not so much the veracity of expert knowledge but its efficiency (Koppl 2018:152). He also says that we need to build reliable systems, a task which can be achieved despite unreliable individual experts (Koppl 2018:219). I share his concern about the privileging of epistemic veracity in much of the literature: indeed we should be concerned about the social outcomes of decisions. Insofar as expert knowledge is a crucial input, we need to ensure our institutions can cope with uncertainty and provide mechanisms for learning, which is to say being able to reverse decisions should they turn out to be wrong.

Different political cultures provide different institutional settings and procedures to deal with the problems, something Sheila Jasanoff has called 'civic epistemology.' Cross-national comparisons are vital for an understanding of these different practices and for their potential to minimize what Koppl calls expert failure. His empirical reference point is mainly the US, with some nods to other Anglophone political cultures, which are all part of the common law tradition. It is doubtful if general insights can be derived from such a narrow sample.

The libertarian spirit which Koppl expresses in sentences like 'pluralist democracy and the rule of experts are incompatible' is perhaps more easily shared by readers from an Anglo-American culture. It sounds strange in the ears of Europeans who live in a different political environment in which expert knowledge is mobilized and articulated. Here, state administration and professional organizations play an even bigger role, and civil society organizations have a deeper involvement with the decision-making process.

Asymmetric information

Experts know more than non-experts in their field of expertise. This is known as asymmetric information and means that experts have the opportunity to exploit the client (Akerlof 1979). In this sense, Berger and Luckmann, and Koppl following them, have a point. But this plays out differently in different institutional settings. Imagine a market-based health system in which doctors get paid through private arrangements (as is the case in the US), through health insurance (as is the case in many European countries), or through a state-run health service financed through taxes (as is the case in the UK).

The US has a problem of access to health services for the poor while the well-off seem to get world leading treatment. Patients able to afford it can choose the best medical experts for treatment. Here, and in Europe there is an incentive for doctors to 'overtreat' patients because they will be paid through private insurance. In the UK, the harsh reality of 'undertreatment' bites, as the national health service does not have the resources to perform even necessary interventions.

The 'doctor scaring the patient' effect plays out differently in these settings. In the US, this would typically occur where the patient has developed a relation of trust which the doctor can exploit. A similar situation obtains in Europe, where the doctor can recommend unnecessary treatment. However, the patient has the opportunity to switch to a different doctor. In the UK, there is no free choice of doctor. Avoiding and delaying treatment is necessitated by the chronic underfunding. Doctors' exhortations are typically restricted to lifestyle choices that would have an effect on the overall cost to the National Health Service (NHS), most notably regarding smoking, drinking, eating. When treatment is advised it is always with a view to be cost-effective in the short run, which may lead to more costly treatment in the long run. Wealthier patients have the option to seek private treatment if only to kickstart their treatment which is then delivered through the NHS.

These examples indicate that Berger and Luckmann's allegory, written in the early 1960s, is less in evidence today. We have not only seen the rise of the well-informed patient but the different institutional pressures that exist in different health systems. There is scope for overselling treatment through fear mongering in market-based systems and for lifestyle exhortations in state-run systems like the UK. But we also note a growing influence of patient groups which have led to the rise of citizen expertise in all countries mentioned.

Where does this leave us? How can we understand the changing landscape of expertise in decision-making? Consider two different accounts. One portrays the rise of the expert society, as *The Rise of the Unelected* by Vibert (2007) does. The other sees the uprising of the audience, *Der Aufstand des Publikums* [the revolt of the audience, R.G.], by Gerhards (2001). I think both are right, and we have to understand the changing landscape as a process which is simultaneously shaped by both forces, one through undemocratic forms of governance based on technocratic expertise and the other through forms and demands of citizen participation and intervention. However, this dialectic plays out differently in different countries and in different domains.

Discussion and conclusion

The review of several frameworks on expertise has revealed some interesting features and some lack of perspective. The interesting features have to do with the limited function of science to provide reliable knowledge for practical purposes and a need to complement the decision-making process with interest groups that originate outside science. There is a growing realization that expertise is ubiquitous. However, this literature also shows to some extent the fixation with scientific and professional knowledge as the standard of reference when talking about expertise, and a neglect of political, institutional, and technological aspects of decision-making.

The argument so far has explored different conceptualizations of experts and expertise. It has advocated a more nuanced framework, emphasizing the proliferation of expertise in the knowledge society, and a mediating role exercised by all kinds of experts, no matter what formal credentials they have. This still leaves open the question of how the expertise relates to decision-making.

There is, of course, the role of the specialist, which combines the cognitive and practical aspects of expertise: the specialist applies a decision without any further mediation. Specialists are partly congruent with professionals, but their social base is much wider. Employees of private and public organizations are able and entitled to implement decisions without further deliberation or external advice.

By contrast, government advisory bodies best exemplify the relational aspect of expertise. Recall Turner's work and his comment on expertise provided to a public authority. He says the problem with advisory bodies is that 'the evidence is not enough to guide practice unequivocally, and that *some sort of fraught step, even a leap, is needed to get from the evidence to practice.* The reason for these institutions is this: the science alone—the "evidence," in this phrase—is insufficient, conflicting, or simply not in a form that can guide practice. But practice is necessary, ongoing, and consequential. So the process of bridging the gap needs to be helped along' (Turner 2010:253, original emphasis).

This brings us back to the sociological insight that knowledge and action are often disconnected, and that action often precedes knowledge. The example of advisory bodies advising high politics is instructive here. Robert Oppenheimer famously suggested that 'there are formidable differences between the problems of science and those of practice. The method of science cannot be directly adapted to the solution of problems in politics and in man's spiritual life' (cited in Shapin 2007:70).

A private person, like a government, knows that expert evidence often is not enough—it may be contradictory, obscure, or inconvenient. A leap is necessary, which all too often is a leap of faith. An advisor can bridge the gap, so to speak. The same logic applies to other social groups requesting expertise, such as companies and civil society organizations. If this is the case, the social attributes of expertise come to the fore. They have been described as defining an issue, reducing complexity, creating trust, and identifying courses of action (Stehr and Grundmann 2011).

These insights, important as they may be, are difficult to digest. The picture painted above, or if you prefer the musical metaphor, the tunes and riffs that we encountered, show us the limits of the idea of science-based decision-making and, by extension of rational decision-making à la Habermas. Official experts as well as decision-makers, are naturally keen to use the kudos of rationality, evidence, and science, especially when under pressure and confronted by uncertainty.

State administrations mobilize the relevant usable knowledge, through their own ministries, through state financed universities and research institutions, and via private firms. The knowledge thus mobilized is often an argument of state power which is dressed up as 'world-class science.' It is supposed to fill the gap between an uncertain world and the need to make a decision. Part and parcel of the process involve the crafting of a convincing narrative. After all, politics is not only a game of power and force but also a game of persuasion (Majone 1989). The preceding chapters indicate that the narratives can either be reassuring, or fear inducing, or both, by depicting different scenarios of the future and by identifying scapegoats for blame.

Narratives are offered by a variety of commentators, some of which are more likely to be heard. Helped along by the media, scientists and governments often occupy the stage, moving quickly and tweaking the story as events unfold. Governing through the manipulation of public opinion in liberal democracies is nothing new, and it is a craft that needs to be mastered. Nevertheless, governments and their scientific advisors often assume that their monopoly position is to be taken for granted. The rise of the commentators in the knowledge society poses a challenge to the monopoly, which is a monopoly of narrative based on knowledge claims. Little wonder that the first line of defence is to belittle other commentators as ignorant and to question their moral status.

The preceding chapters have also shown how science policy advice can be understood. Several authors emphasized the advantage offered by diverse teams. Advisory groups that are epistemologically diverse are more likely to deal with a problem in a way that leads to 'success,' i.e. a better understanding of the problem, perhaps even a better definition of solutions and strategies to implement them. Homogenous groups are less likely to achieve this. In addition, the open discussion of relevant issues, where arguments are presented in public, is likely to increase the quality of decision-making and to enhance trust in decisions. Reaching pragmatic compromises based on the interests of a multiplicity of relevant groups is perhaps the best prospect we have when it comes to making decisions under uncertainty and conflicting values. However, we should not be carried away by this. There are severe limits as regards representation and inclusion of a multitude of perspectives, and even the best pragmatic arrangements will be temporal and provisional.

In part II, I will discuss various examples in the light of my conceptual framework. The examples are the cases of the IPCC, Covid-19, and three professional fields of expertise: medicine, journalism and law.

Part II

Chapter 7

The IPCC

A Chameleon of Expertise

In public discourses about climate change across the world, the IPCC is arguably the organisation which attracts most attention. It is ubiquitous in the climate change discourse and much has been written about it. I put forward the thesis that it is unique as an organization that provides expertise, but it needs to be better understood.

The IPCC defines its goals as follows: 'Created in 1988 by the World Meteorological Organization (WMO) and the United Nations Environment Programme (UNEP), the objective of the IPCC is to provide governments at all levels with scientific information that they can use to develop climate policies. IPCC reports are also a key input into international climate change negotiations.'

It is often said that the IPCC is an 'authoritative voice in climate science relevant for policy' (Sokona et al. 2015), or that it is the 'primary scientific authority to policy-makers' (Paglia and Parker 2021). Its reports 'are widely regarded as the most important and authoritative publications on a global scale' (Howarth and Painter 2016). A practitioner in the field allegedly called them 'the equivalent of the King James Bible on climate change' (quoted in Painter 2015:3).

Some authors think the IPCC is exaggerating the risks of climate change; others think it is downplaying them. Some think it is an honest broker; others think it is an issue advocate. Some think it is scientific body; others hold it is a boundary organization between the worlds of politics and science. Of course, the IPCC could be all of these, in different circumstances. Even if this were true we still need to understand what the IPCC is, what it does, and how it relates to the issue of expertise. I will try to untangle the conceptual maze by applying my typology which provides the analytical tools to understand its role and performance. I put forward the thesis that the IPCC is one of the most important commentators on climate change, a role it shares with many other voices. I contend that the IPCC is a chameleon. It combines the roles of science arbiter, honest broker, stealth advocate, and commentator. But it scarcely delivers policy advice in the conventional sense.

The chapter has the following structure. I will first summarize relevant approaches from the literature, then present the genesis and function of the IPCC,

DOI: 10.4324/9781003320227-10

including criticisms of its operation. I will then spell out what kind of expertise the IPCC has to offer and what kind of narrative prevails in public discourse.

As this chapter draws on my own research, and my evaluation of other contributions in this field, it is going to produce a unique tune. There will be a constant underlying beat, reflecting the types of expertise I have identified. Then there is a background loop, reflecting what I have to say about other interpretations on the basis of my own tune. All in all, it is going to produce a complex soundscape which differs from other chapters in this book.

Meanings of expertise

As the reader will be familiar with my argument by now, I provide only a very brief recap distinguishing several meanings of expertise. Authors have defined experts as scientists (Collins 2014; Pielke Jnr. 2007), as professional specialists (Abbott 1988; Weber 2012), and as citizen experts (Funtowicz and Ravetz 1993; Wynne 1996). Scientific expertise has been further differentiated into the roles of science arbiter, issue advocate, or honest broker (Pielke Jnr. 2007).

Not only persons provide expertise, but organizations and networks do as well. In fact, many expert roles in modern society can only be performed as part of a larger socio-technical network, as the cases of pilots or doctors exemplify (Eyal 2019). Scientific advisory bodies are well-researched examples of organizational forms of expertise (Jasanoff 1990; Lentsch and Weingart 2009).

Sometimes the relation between the expert and the client who requests expertise is emphasized (Stehr and Grundmann 2011). The client may also seek advice from the expert when facing uncertainty and conflicting values (Grundmann 2017). The client typically wants to know what the best course of action would be. To this list we need to add a group of experts who do not implement decisions, and who do not advise clients: I call these experts pundits, or commentators. They range from public intellectuals to sports commentators, to financial analysts.

Much previous research on the IPCC has focused on the science advisory aspect. In these contributions the IPCC is described in terms of co-production, boundary work, issue advocacy, or post-normal science (Beck and Mahony 2018; Guston 2001; Jasanoff 2004; Mahony 2013; Miller 2004; Pielke Jnr. 2007; Ravetz 2011; Hoppe et al. 2013; Saloranta 2001; Turnpenny et al. 2011). These contributions have all produced important insights into the role of the IPCC at the science-policy interface. However, more needs to be said. I start with the issue of science advice before delving into the history of the IPCC.

Summarizing various case studies of scientific advisory bodies across several countries, Lentsch and Weingart (2011:354) provide four characteristics for a scientific advisory organization:

- it is an institutionalised (or statutory) body;
- it performs its internal operations independently, i.e. free from external influences, particularly from its clients;

- its activities are grounded in or based upon scientific expertise; and
- its mandate essentially comprises to advise the legislative or the executive, i.e. the government or governmental departments, on science-related policy issues.

While the first and third items apply to the IPCC, the second clearly does not, because the client is involved. It is an intergovernmental body which means that national governments are involved in authoring its reports: governments nominate the authors for the report, and they approve the Summary for Policy-makers (SPM) document in a line-by-line procedure. This makes the IPCC a very special organization, even within the world of policy advice. It is a body that marshals, assesses and thereby produces knowledge about the climate system, and reports to the United Nations.

But it does not produce advice which means the fourth criterion does not apply either. The IPCC does not advise national governments, nor does it advise the UN in the strict sense (although its reports are referred to in negotiations of the Conference of the Parties (COP)). National governments have set up their own advisory bodies and pursue different, sometimes opposing policy options (Jasanoff 2011a; Grundmann 2007; Skolnikoff 1997).

Pielke Jnr. (2007:141) picks up on the question of policy advice, saying that the IPCC 'was formed to provide guidance to policy-makers on climate change,' and 'by design does not discuss policy options, yet the IPCC report and its representatives often serve as Stealth Issue Advocates because some policy options are discussed and others ignored.' His example is the preference for mitigation over adaptation policies. Providing guidance could of course mean many things, and Pielke's view is that science advisors should outline a range of policy options. Honest brokers are there to expand the range of options.

One might say, as Pielke (2007:151) himself acknowledges, that 'decision-makers look to political advocates for the insight into the significance of science for action, in effect creating a world where almost all science is filtered through existing special interests.' This would be an argument which draws on the legitimizing function of knowledge, as I described above as *knowledge politics*. But Pielke does not follow this line of thought. Instead, he presents a normative model: 'A better alternative is for Honest Brokers of Policy Alternatives to take some responsibility for addressing the significance for policy of scientific results' (Pielke 2007:151). Before discussing normative questions, I want to address the political context in which policy makers elicit, and use scientific claims for justifying political action.

Edenhofer and Minx (2014) address the problem in a slightly different way to Pielke. They distinguish between a world of science and politics, where the science provides the maps for policy-makers who in turn use them for navigation. Using the metaphors of mapmaking and navigating, scientists are expected to stick to the role of mapmaking, while politicians do the navigating: 'From such common understanding, the IPCC can further inform international climate policy without prescribing and predetermining future negotiations.'

Like Pielke they give the scientists the task to provide 'guidance' to politicians, but unlike Pielke they do not call for an expansion of policy options. The mapmaking metaphor is instructive here: what is not on the map is not relevant for navigation. The reverse is also true: putting some things on the map will ensure that policy-makers are aware of them, even if they choose to neglect them. And there is always the risk that things which are not shown on the map may wreak havoc on your seafaring journey.

Placed in this web of metaphors the IPCC has some power to tacitly define priorities. In line with Pielke's notion of stealth issue advocacy, and the concept of agenda-setting in political science (Schattschneider 1960), we note that what people think about, and how alternatives are defined, are prime tools in the exercise of political power. This idea is developed by Beck and Mahony who claim that 'the IPCC is an incredibly powerful actor in climate politics' (Beck and Mahony 2017:313). As they show with the example of bioenergy with carbon capture and storage (BECCS), the inclusion of this technology in scenarios was crucial to arrive at net-zero emission targets. They argue that these scenarios are not mere academic exercises, but performative: Such 'pathways and scenarios do not just represent possible futures, but also help to bring certain futures into being.'

BECCS has been given this prominence as it appeared to IPCC authors as a technically feasible and economically optimal solution to the problem of achieving the 2°C warming limit. The authors are critical of this approach and call for a more inclusive approach towards options: 'The IPCC could open up consideration of alternative technological possibilities, of future pathways of climate policies, and alternatives to current policy orthodoxies' (Beck and Mahony 2017:312). While this call for openness and inclusion is commendable, the question remains if it is realistic to envisage the IPCC in such a way as to deliver these aims.

In order to answer this question, we need to consider the nature of expertise offered by the IPCC. I will do this via a historical reconstruction of its genesis and role before summarizing criticisms of its practice. The main result of this exercise is that the IPCC offers highly visible commentary to a world audience but does not provide policy advice in the usual meaning of the term. It also combines the role of science arbiter with stealth advocacy, but does not fulfil these roles in a transparent way.

Genesis and function of the IPCC

The IPCC was founded in 1988 by two UN organizations, the WMO and UNEP. In the mid-1980s WMO, UNEP, the International Council of Scientific Unions (ICSU) and the United States led efforts to establish such a body (Agrawala 1998). This was done through a process that went in parallel but separate to the setting up of the United Nations Framework Convention on Climate Change (UNFCCC). The convention was signed in 1992 and stated as its ultimate objective the 'stabilization of greenhouse gas concentrations in the atmosphere at a level that would prevent dangerous anthropogenic interference with the climate system' (UNFCCC 1992:Art. 2). What exactly the word 'dangerous' meant in this context was not spelled out and it took several years before a political consensus emerged that provided an answer (Randalls 2010).

The United Nations General Assembly, at its 70th plenary meeting in December 1988, defined the IPCC's mandate as follows: to 'provide internationally coordinated assessments of the magnitude, timing and potential environmental and socio-economic impact of climate change and realistic response strategies' (United Nations 1988). It also requested the WMO and UNEP to utilize the IPCC to undertake a 'comprehensive review and recommendations with respect to … possible response strategies to delay, limit or mitigate the impact of adverse climate change.'[1] Three working groups (WGs) were established: Science (WG1), Impacts (WG2), and Response Strategies (WG3).

The US was keen to play a leading role in setting up the IPCC and took the initial lead of WG3. The US probably wanted to gain control over a potentially unruly public debate about the size of the threat, curbing environmentalist campaigning with the support from climate scientists (Bodansky 1993:465). Scientists on the other hand welcomed the prospect of speaking with one voice which would make political responses timelier and more effective, or so they hoped (Boehmer-Christiansen 1994a, 1994b; Haas 2004; Hoppe and Rödder 2019; Siebenhüner 2002; Skodvin 2000). Communicating a consensus message was assumed to be instrumental in this process. The legacy of this argument has cast a very long shadow in the climate discourse.

In addition, the intergovernmental structure of the IPCC gave developing countries an opportunity to participate, avoiding the problem of mistrust of science developed by rich countries of the global North. As Bert Bolin, the chair of the IPCC in its early years put it, 'right now, many countries, especially developing countries, simply do not trust assessments in which their scientists and policymakers have not participated. Don't you think global credibility demands global representation?' (cited by Schneider 1991:26).

Developing countries were also suspicious of the potential unwelcome messages from the IPCC. They wanted to ensure that its focus of activity was restricted to the science, where they demanded more participation. Political messages should be avoided, as Miller (2004:60) observed: 'Developing countries, who had actively lobbied throughout the first two years of the IPCC for greater participation, eased their criticism of the organization with the apparent decrease in the Panel's ability to make political decisions.'

Until 1992 WG3 came close to provide political advice, and was the most politicized of the three working groups. But in a significant move in 1992, WG3 was reorganised: the old WG3 was merged with WG2 and a new WG3 was set up to deal with socio-economic and other cross-cutting issues. In the

Second Assessment Report, published in 1995, the remit of WG3 was *Economic and Social Dimensions of Climate Change*, and from the third report (2001) onwards it became *Mitigation of Climate Change.* The term 'response strategies' had disappeared altogether which is to say that the post-1992 mandate of WG3 eliminated the element of policy advice.

The IPCC says it is 'providing policy-relevant but not policy-prescriptive information.' This formula was not used until the Fourth report was published in 2007.[2] No mention of policy-relevant options was made in the first report published in 1990. The second report from 1995 only mentioned in passing that it is an assessment of the knowns and unknowns, but 'not a prescription for policy implementation.' The Third report did not mention the formula. Only the Fourth report carried a foreword which states: 'The IPCC does not conduct new research. Instead, its mandate is to make policy-relevant – as opposed to policy-prescriptive – assessments of the existing worldwide literature on the scientific, technical and socio-economic aspects of climate change.' The formula of being policy relevant but not policy-prescriptive was repeated in the Fifth report of 2013. More about this formula later.

Still, the IPCC does influence public opinion and politics through its reports. The assessment of the scientific literature is the process through which this influence becomes evident. Assessing the published studies means to foreground specific studies, and to background other studies, thereby defining issues for attention. This framing activity (Entman 1993) includes definition of the problem, causal analysis, and future projections. Other aspects of framing are less evident: the IPCC does not apportion blame, and does not advocate specific remedial action.

It is in this respect only that one can describe the IPCC as having power (Beck and Mahony 2017), even being a monopoly (Tol 2010). It is in this respect that the IPCC can be said to draw the map and to set the agenda. In framing the problem of climate change, the IPCC and the commentators who speak in its name have occupied the position of defining policy alternatives. As we shall see, this is mainly done through scenario building, contrasting 'business as usual' futures to intervention paths. As I argued above, this makes politicians nervous as the IPCC potentially takes away their natural prerogative. In other words, the IPCC has a narrowly circumscribed role as being 'policy-relevant, not policy-prescriptive,' but it has somehow managed to implicitly favour some options over others. To repeat, it does this not through traditional channels of science advisory systems but through its comments to a world audience, and thereby to international institutions and national governments.

The prominence and influence of the IPCC is surprising given the limited role envisaged initially in the institutional framework of the global climate regime. In the 1992 Framework Convention it is not even mentioned as a permanent body. The bodies specified by the convention are the COP, the Secretariat, and two subsidiary bodies: the Subsidiary Body for Scientific and Technological Advice (SBSTA) and the Subsidiary Body for Implementation (SBI). The COP is the supreme decision-making body under the Convention.

The IPCC only appears in Art. 21(2) where Interim arrangements are detailed. The reference to the IPCC is as follows:

> The head of the interim secretariat ... will cooperate closely with the Intergovernmental Panel on Climate Change to ensure that the Panel can respond to the need for objective scientific and technical advice. Other relevant scientific bodies could also be consulted.[3]

This means that the IPCC was not envisaged to be an important, permanent body within global climate change governance, let alone a central one. As the quote makes clear, other relevant scientific bodies could be consulted when scientific and technical advice is sought. As Bodansky (1993:536) commented: 'This limitation reflects the concern of some states that providing advice directly to individual parties could be intrusive.' In other words, it is SBSTA that will provide advice to the COP (based, inter alia, on the IPCC reports). This advice needs to be in a form that is digestible by the COP. It needs to be digestible with regard to form and substance: it needs to contain only such stipulations that are already on the radar of the governments of the world.

The involvement of governments is even more direct in the writing of the reports. The Summary for Policymakers documents are agreed upon line-by-line with government representatives. The expert advice provided to the COP is thus two steps removed from the science: first by the inter-governmental editing process of the Summary Reports, and secondly by the fact that scientific information will be received by, and fed into the COP, by the SBSTA. SBSTA is free to elicit and use advice from other bodies than the IPCC. This means that the IPCC's position in the knowledge regime of global climate governance is potentially precarious.

In 2010, after Climategate and several errors in the 2007 IPCC report received high levels of publicity and scrutiny, chairman Pachauri was not invited to participate in the opening ceremony of the 17th COP. The participation of the chairman of the IPCC at COP meetings had been customary since COP1. Instead, Pachauri was told to attend the meeting of SBSTA. After some behind-the-scenes negotiations he was, after all, allowed to address the COP assembly (Gutiérrez et al. 2011). This little episode illustrates the precarious position of the IPCC. It had been invited to speak at COP plenaries, but this was a discretionary act, not a constitutional right.

As one commentator pointed out: 'Although these events may have been unrelated, the criticism directed at the IPCC since 2009 has undermined the organization's scientific authority and shaken its position in the climate field. This loss of legitimacy has the potential to impact those closely coupled with the organization, an effect the UNFCCC would want to avoid, particularly when it is dealing with questions regarding its own centrality to international climate action' (Hughes 2015:97).

What kind of knowledge does the COP expect from SBSTA? In the early years a major task had been identified in the need to establish a shared system of metrics. This meant to set up national greenhouse gas inventories, and agreed-upon methods to measure emissions, and emission reductions. This task proves to become especially relevant after the Paris Agreement as countries engage in a pledge and review process where reliable and uncontroversial data is needed. The task is conducted by a Task Force on National Greenhouse Gas Inventories (TFI) which aims to develop and refine an internationally agreed methodology and software for the calculation and reporting of national greenhouse gas emissions and removals, and to encourage the use of this methodology by parties to the UNFCCC.

Developing reliable and agreed-upon metrics has extended to other aspects related to climate drivers and their impacts. An important metric is the global warming potential (GWP) which makes different greenhouse gases comparable with regard to their effects. GHGs are expressed as CO_2 equivalents and calculated accordingly. Another set of metrics pertains to the estimation of temperature changes related to a specific unit of emissions. Because emissions correlate with global temperature rise, global emissions budgets ('carbon budgets') have been calculated which would provide guidance for policy-makers to stay below dangerous levels of warming (to be precise, there is a difference between CO_2 and GHG budgets, see Peters 2018; for the history of the concept see Lahn 2020). Still another metric relates to impacts stemming from the rise in temperature. These efforts of standardisation have enabled scientists to translate different metrics into each other, and to combine basic scientific insights in a narrative for public consumption. Most of these metrics are based on physical science.

We can call this the epistemic core of the IPCC. It can be seen at work in a powerful narrative about a planet in danger. Computer simulations, images, and rhetoric have been used to illustrate the risks of increased GHG concentrations, and these are widely used in public discourse. The physical sciences have come to a shared way of expressing key metrics which are translatable between each other while parts of the biological, political and social sciences are hesitant to develop and use such metrics (with economics being the exception, more on this below).

There are various methods and tools to establish knowledge in this field, with General Circulation Models (GCMs) occupying a central position. These are three-dimensional representations of the atmosphere, oceans and land. Climate scientists perceive GCMs to be top in a hierarchy, the best way to perform 'good science' (Shackley et al. 1998:193). The models are based on numerical equations and rely on computers. Over the past decades the computing power to run these models has increased manifold. These models are 'global knowledge infrastructures' (Edwards 2010). The metrics are the product of a centre of calculation, a scientific-technical-political infrastructure which produces lists of shared facts (Bowker and Star 1999; Edwards 2010; Latour 1987).

An early but still fundamental metric used by climate scientists is the effect a doubling of CO_2 concentrations has on the global average temperature, compared to pre-industrial times–a specific instance of climate sensitivity. It can be translated into a measure of radiative forcing expressed as additional energy absorbed per square meter. This measure is used to estimate what the effects of increased energy absorption on global average temperatures will be. This estimate is not a precise single number but a range, from 1.5°C to 4.5°C. Over the past decades, climate science has made no progress in narrowing down this range, or to come up with a single number.

For its Fifth Assessment report (AR5) the IPCC has introduced a new way of projecting future developments, Representative Concentration Pathways (RCPs). These provide projections about the CO_2 concentration at the end of the century and the resulting warming (Rogelj et al. 2012). They range from 400 ppm CO_2 (and a radiative forcing of 2.6 W/m^2) to 1370 ppm CO_2 (and 8.5 W/m^2). Temperatures would rise between 1°C and 3.7°C (median values). Sea levels would rise between 26 cm and 82 cm at the end of this century (van Vuuren et al. 2011; more recently, in a special report on Oceans and the Cryosphere, the IPCC (2019) projects the maximum sea level rise to 1.1 m). A paper from 2018 introduced an even lower, 1.9 RCP in response to the 1.5°C in the Paris Treaty (Rogelj et al. 2018).

The IPCC did not assign probabilities to any of these pathways but scientists assume that the lowest and highest RPCs describe extreme cases. RPC 1.9 and 2.6 are only plausible with BECCS or other *geoengineering* instruments. RCP 8.5 is a *worst case scenario*, assuming that no climate policy is enacted. It is often, and inappropriately, presented as *business as usual* (Peters and Hausfather 2020).

In the Sixth Assessment Report yet another method has been used to differentiate between five different pathways, 'that cover the range of possible future development of anthropogenic drivers of climate change found in the literature.' In this report, the IPCC lists scenarios 'with high and very high GHG emissions (SSP3-7.0 and SSP5-8.5), scenarios with intermediate GHG emissions (SSP2-4.5) and CO2 emissions remaining around current levels until the middle of the century, and scenarios with very low and low GHG emissions and CO2 emissions declining to net zero around or after 2050, followed by varying levels of net negative CO2 emissions (SSP1-1.9 and SSP1-2.6).' (Masson-Delmotte et al. 2021).

One might think that these are highly technical devices and have little bearing on the climate change discourse. This would be quite mistaken. These scenarios relate to the double narrative structure of raising fear and reassurance. They therefore play an important role in communicating the potential risks of climate change, depending on which path world society takes in the coming years and decades. Catastrophic developments are on the cards, but we could be OK if we do the right thing. The ubiquitous five minutes to midnight image (Geden 2018) captures both elements in a powerful metaphor.

To sum up the argument so far: the IPCC has the role of *science arbiter*, evidenced by the assessment of the available literature. It is the central part of a knowledge infrastructure which relies heavily on systems of unified metrics. It also has the role of *stealth advocate* through scenario development, and it has the role of *commentator* by offering a narrative for public consumption. Its official mandate has changed from providing response strategies to being policy relevant but policy neutral. This means that the IPCC is prevented from providing practical knowledge for policy makers. In this sense, it does not qualify as an expert body in the usual meaning of an advisory body (see Edenhofer and Minx 2014 for a defence of this restriction).

If this is a watering down of the remit and function of the IPCC, it has also assumed a much larger role compared to what was laid out initially. The IPCC has become the central reference point in global discussions about climate change, despite the fact that its statutory role is much more circumscribed. It has also gone beyond the task of assessing the state of the art in climate science: it has produced a battery of metrics, tools, and approaches which have come to be used in the field of climate research and in public debates. The IPCC has produced elements to frame the problem of climate change and became a common reference point for discussions about climate change.

The IPCC has, of course, attracted a lot of criticism. By this I do not primarily mean voices who are sceptical about the science claims. Instead, I want to focus on an alleged bias which pertains to geographical and professional/disciplinary representation.

Criticism of the IPCC: Geographical and professional bias

There are two biases in the IPCC which have been identified over and over again: the dominance of Northern countries, and the dominance of natural science disciplines (see, e.g. Corbera et al. 2016; Hulme and Mahony 2010). While the uneven geographical participation potentially challenges the legitimacy of the IPCC process, the disciplinary biases restrict the kind of knowledge that is produced, the kind of questions that are asked, the kind of problem that is constructed, and the kind of solutions that are envisaged.

The following quote expresses a common criticism: 'Northern domination of science globally relevant to climate change policy and practice and lack of research led by Southern researchers in Southern countries may hinder development and implementation of bottom-up global agreements and nationally appropriate actions in Southern countries. The divide may impact most on least developed countries (LDC) and small island developing states (SIDS), which are the most vulnerable to climate change but contribute least to relevant research' (Blicharska et al. 2017:21).

The authors call for more cooperation and programmes that would facilitate inclusion of researchers from the Global South. However, the problem is perhaps much deeper, as Hughes and Paterson (2017) show with regard to authorship patterns in Working Group 3 in AR5. Based on a social network analysis they show how a group of 44 highly connected and highly cited researchers are dominating the work of WG3. A few research institutes in the global North serve as the organisational springboards for influence.

Turning to disciplinary hierarchies, the science element of IPCC assessments from WG1 gets more of political and public reception compared to the reports from WG3. This is an intentional design feature, as national governments do not want to receive recommendations from the IPCC. The formula of being policy relevant but not policy prescriptive has exactly this function. The policy-relevant knowledge the IPCC delivers is very general, embedded in a bureaucratic and diplomatic framework, and tied to the epistemic core of WG1.

WG1 publishes results from model calculations, emphasizing observed changes in physical parameters, the causes of observed change, and projections into the future. The main elements in WG3 reports are economic, linking economic value to physical parameters. There is a near-complete absence of the human and societal dimension (Hackmann et al. 2014; Heymann 2018; Holm and Winiwarter 2017; Weaver et al. 2014).

Yearley (2009) notes the absence of Sociology in the IPCC, Fiske et al. (2014:20) the marginal role of anthropology: 'Anthropology is uniquely placed to question the meaning of climate change, how it is constructed by different parties including science and political actors, and how it is maintained. An important role of anthropology to "problematize the problem" of climate change and to challenge the basic cultural assumptions that underlie IPCC approaches to climate change.'

Another criticism pertains to the absence of an assessment of the findings over time, and lack of any systematic review of the literature: 'A clear evolution in the findings over time is lacking, which supports our claim that knowledge has failed to accumulate; the findings do not reveal a systematic assessment of the literature; and the statements are typically overly generalized or purely descriptive in nature with little policy relevance' (Minx et al. 2017:255).

The importance of this criticism cannot be overstated. The growth of climate-related research is exponential and so rapid that clearly defined procedures need to be in place to assess its content, unless the reviewers want to be accused of cherry-picking. To put the growth in knowledge into perspective, a doubling of climate change papers occurs every 5–6 years, while overall scientific publications double in 24 years (Haunschild et al. 2016).

Given that the assessment of the scientific literature is the key task for the IPCC it is remarkable that it has not made prominent use of systematic reviews in its efforts to assess the scientific knowledge base. Systematic reviews are regarded as best practice in many fields of research, especially in medical research. Its absence in the IPCC is very odd. Here is a description of the problem:

> The first important aspect to note is that not only has the body of literature on climate change grown, but so has the number of unique citations made in IPCC reports: from AR1 to AR5 there has been a steady increase

> from about 1600 to about 31,000 citations. This is not only due to the increased length of each report, but also to an increase in citation density (citations per written page), from 1.6 citations per page in AR1, to 6.4 citations per page in AR5.
>
> (Minx et al. 2017:253)

The absence of self-reflection is noteworthy. Reports from institutionalized expert committees usually give an account of how the findings of a new report relate to the previous report, a practice well known from economic advisory panels. This allows the client, and the audience, to understand the reasons for a new assessment, and perhaps a new policy recommendation. Because the IPCC does not have a government as client, and must not recommend policies, this blind spot is not surprising. It nevertheless needs to be covered, as the audience expects such context. This is done by individual commentators. These commentators are authors of the IPCC report itself, or other climate scientists, but in principle everyone: journalists, activists, politicians, business leaders, or professional organizations.

The argument so far has provided evidence for the claim that the IPCC is officially prevented from developing response strategies to climate change risks but provides a mix of science arbitration and commentary which are communicated through scenario-based narratives. The IPCC is dominated by scientists from rich countries who have the power to define issues and to frame options. The role of science arbiter is performed in a non-systematic way, without an account of how the assessment itself has changed over time. In what follows I will examine the nature of these IPCC communications to the world audience in more detail.

IPCC and public discourse: The narrative of climate change

The IPCC does not have what one could call an elaborate corporate PR strategy. It seems to leave the communication of its findings to the chairs of the three working groups, of which the leader of WG1 is in most demand from media enquiries. This has to do with the disciplinary hierarchy discussed above, which is reflected in the timing of the release of IPCC reports. WG1 comes first, and the world's leading media outlets cover its report extensively. By the time reports from WG2 and WG3 are published, the news agenda usually has moved on to different issues. Sometimes there are press releases which include short summaries, video clips, or press conferences.

However, climate scientists are not professional communicators, and the choice of topics and story lines is done in an idiosyncratic way. Hollin and Pearce (2015) give an example of an official press conference which was held 2013 in Stockholm after the publication of WG1 report. The speakers comprised the heads of the UN, WMO, UNEP, and IPCC. The big talking point was the level of certainty about AGW, which was a hot topic in the aftermath of public debates about an alleged 'pause' in global warming in the run-up to the publication of the report.

Boundary work

Such press conferences are an exception. On a daily basis the PR of climate change science is performed by commentators who sometimes, but not always, have an official affiliation with the IPCC. There are also other commentators, some of them climate scientists, who have no affiliation with the IPCC but purport to speak on its behalf. Over the years their role has increased, arguably through the availability of social media platforms. Then there are other scientists, including climate scientists, who are critical of IPCC reports. In a way this commentary role is a free-for-all. For the interested public the question arises, who should we believe? Someone who has been involved with the drafting of a chapter, perhaps in a leading role? Or a highly reputed scientist who hasn't? There is no reputational scoring tool that could be used, and most observers select statements and narratives they believe in, find persuasive, or useful in a political debate.

Consider this example as *pars pro toto.* Dozens of climate scientists wrote in the Wall Street Journal:

> Do you consult your dentist about your heart condition? In science, as in any area, reputations are based on knowledge and expertise in a field and on published, peer-reviewed work. If you need surgery, you want a highly experienced expert in the field who has done a large number of the proposed operations. You published 'No Need to Panic About Global Warming' (op-ed, Jan. 27) on climate change by the climate-science equivalent of dentists practicing cardiology. While accomplished in their own fields, most of these authors have no expertise in climate science. The few authors who have such expertise are known to have extreme views that are out of step with nearly every other climate expert.
>
> (Trenberth et al., cited in Ramírez-i-Ollé 2015:395)

Above I mentioned the important sociological concept 'boundary work', coined by Tom Gieryn. He mentions the mechanisms of monopolization, expansion, expulsion, and protection of autonomy as part of boundary strategies employed by scientists. While Gieryn's notion is useful to understand these attempts, the case of climate change also shows that boundary work undertaken by climate scientists from the 'core group' (Collins) is not always successful. To be sure, the above mechanisms are evident as part of rhetorical strategies. But the climate problem has become so all-encompassing that attempts to keep it under the wraps of the IPCC, or of 'official climate science' have become futile.

Narratives of climate change

There are numerous narratives in public discourse on climate change. The social science literature distinguishes different types of narratives. An influential approach, Narrative Policy Analysis, holds that a narrative consists of four elements: setting, plot, characters, and solutions (Jones 2014; Jones and McBeth 2010).[4] The setting is a commonly accepted description of the issue, the plot contains causal and moral claims, the characters are usually heroes, villains, and victims. Each narrative has a moral, and develops policy solutions.

Another narrative analysis is presented by Smith and Howe (2015), who describe climate change as a social drama. Its elements are anthropogenic climate change as a very serious threat to the entire planet. Economic, social, political, and biological systems are in trouble. There is little time to do something about it, while the public is not really involved, the authors maintain. This is partly due to the technical ('scientistic') framing of the issue by scientists who are poor communicators, holding naïve assumptions about the policy process. It is also due to an environmental movement following their heed to 'an uncritical belief in the power of the scientific worldview' (Smith and Howe 2015:120).

In her book *Scientists as Prophets,* Lynda Walsh (2013) casts climate scientists as prophets who struggle with the communication of uncertainty. They use rhetorical devices, such as telling a story of increasing certainty about scientific aspects, in the hope that more certainty would make climate policies easier to achieve. Another device is to represent the potential impacts as dire warnings about the future, as a jeremiad pointing to the sinners' failed ways. She quotes from the Fourth Assessment Report which stated that 'hundreds of millions of people [would be] exposed to increased water stress' and 'up to 30% of species at increasing risk of extinction' (IPCC AR4 Synthesis Report 2007:10).

Prophets cannot be effective through monologue. In order to be heard and followed, they need supporters who propagate the message, and these alarming messages are duly picked up and spread by activist groups. But neither can the text serve as a literal source for engagement. The IPCC reports are often, only half-jokingly, labelled as The Bible of Climate Science. Christians are divided on the issue if believers should read the Holy Text themselves. After all, these texts are difficult to understand, and you might get the wrong idea if you do read them on your own. During the late Medieval period the Bible became translated into vernacular languages in Europe. For example, in Germany, Luther's tracts were distributed in 300,000 printed copies between 1518 and 1520 alone. This provided an opportunity for educated lay people to read the text unsupervised. It is only a slight exaggeration to say that the Catholic church, as supervisor of the faith, lost control of the narrative and disintegrated. Might the rise of social media since the Fourth Assessment Report do the same to the IPCC?

I emphasized the role of commentator which the IPCC plays and that other commentators are visible, too. Some offer vernacular versions of IPCC reports;

others develop heretical views. The IPCC does not have a communication monopoly, nor does it have a communication strategy. Because its reports have many layers and messages, commentators of all persuasions use it as a reservoir for evidence they can use to advance their issue. While in the past the IPCC has been criticized for being too dramatic and activist, especially after the publication of the Fourth report (Pielke 2010), more recently it has been criticized of being too conservative, underestimating the risks posed by climate change (Oreskes et al. 2019).

However, increasingly statements about climate change are made without a link to the IPCC. We hear often terms like 'climate emergency' or 'climate chaos,' suggesting that the IPCC would subscribe to these. In public discourse such terms, and many other claims are widespread, without the IPCC being able, or willing, to take a stance. It is individual scientists who produce comment, and one cannot assume that the statements of those who are deeper affiliated with the IPCC automatically carry more weight than those who are not. This has led to a democratization in the climate discourse, in which the IPCC text has a phantom presence.

The IPCC speaks to the world very much like an oracle. It reminds us of the change we have brought about in terms of emissions, temperature rise, sea level rise, extreme weather, and biological stress. It also seems to suggest that it will get worse, and it tells us what dire consequences our daily lives may have if we do not repent. This genre of communication is known from other prophetic examples, as Walsh (2013) argues. The IPCC performs this role based on computer models and scenarios, which have replaced crystal balls and other instruments from previous times. Parts of the audience who are interested in detailed information probably expect more certainty than the IPCC can deliver. Scientists are not prophets, 'but it is also true that their audiences keep saying they are, and keep working with them toward certainty', as Walsh (2013:8) put it.

According to Walsh this meme of looking for a prophet is deeply engrained in human cultures, at least since the Hellenistic period, and explains why we always come back for more of the same story. She writes: 'The fear that the jeremiad mongers casts the audience back on the prophet as their only hope for more information or intercession with the divine judge' (Walsh 2013:181). We come back, time and again, to listen to, and watch more bad news about the future of the planet, and our lives within it: '[T]he jeremiad's indictment taps into the audience's repressed guilt for having transgressed covenant values, while its vividness confirms the prophet's privileged access to knowledge.' Recall Berger and Luckmann's (1966:105) statement that 'the general population is intimidated by images of the physical doom that follows 'going against doctor's advice'; it is persuaded not to do so by the pragmatic benefits of compliance, and by its own horror of illness and death.'

But remember: expert narratives not only have the element of fear, but also of reassurance. We will see how the glimmer of hope is kept alive, and that claims to prevent catastrophe are made, if people follow the experts' counsel.

IPCC headlines: From bad to worse to hope?

Here I survey the headlines of IPCC reports over time. There seems to be a drama unfolding as we go from the first report, published in 1990, to the latest published in 2021. While the dramatic messages are on the increase, there is also hope on offer. This makes the overall message ambiguous, and different interpretations of the same text become legitimate in public comments and debate.

The first report pointed to a link between increasing emissions from human activities and additional warming of the Earth's surface but said that the observed temperature rise was not bigger than the natural climate variability.

The second report, published in 1995 stated that 'The balance of evidence suggests a discernible human influence on global climate.' It also stated that there are still many uncertainties. The third report was published in 2001. It stated that since the mid-20th century, most of the observed warming is 'likely' (greater than 66% probability, based on expert judgement) due to human activities. It also said that 'Projected climate change will have beneficial and adverse effects on both environmental and socio-economic systems, but the larger the changes and the rate of change in climate, the more the adverse effects predominate.' While this was cautiously worded, the report also included the more ominous sentence, 'When rapidly forced, non-linear systems are especially subject to unexpected behavior.'

The fourth report from 2007 stated that 'Warming of the climate system is unequivocal' and that most of the global average warming over the past 50 years is 'very likely' (greater than 90% probability, based on expert judgement) due to human activities.' Impacts of climate change are expected to increase 'due to increased frequencies and intensities of some extreme weather events.' Due to the long lifetime of CO_2 the warming and sea level rise 'would continue for centuries even if GHG emissions were to be reduced sufficiently for GHG concentrations to stabilise.' There was a stark warning that 'unmitigated climate change would, in the long term, be likely to exceed the capacity of natural, managed and human systems to adapt.'

The fifth report came out in 2013. It pointed to the unprecedented concentrations of GHGs and stated that 'Warming of the climate system is unequivocal, and since the 1950s, many of the observed changes are unprecedented over decades to millennia.'[5] The human influence was stated as 'extremely likely (95–100% probability)' to have caused the warming between 1951 and 2010. It also said that the current trajectory of global greenhouse gas emissions is not consistent with limiting global warming to below 1.5°C or 2°C, relative to pre-industrial levels. AR5 also introduced 'calibrated language' to evaluate and communicate the degree of certainty in its findings. It relies on two metrics, *confidence* and *likelihood.*

Confidence levels range from very low to very high and are based on judgements about the validity and robustness of findings. Likelihood refers to probabilistic statements, based on data, model results, or expert judgement.[6]

The sixth report was published in 2021. Compared to AR5 the headline statements from WG1 SPM are more outspoken, and more dramatic. Not only are the changes in the climate system, and the human causes 'unequivocal.' These changes are 'widespread and rapid.' The scale of the changes is 'unprecedented.' Furthermore, 1.5°C and 2°C warming 'will be exceeded during the 21st century unless deep reductions in carbon dioxide (CO_2) and other greenhouse gas emissions occur in the coming decades.'

At the same time the report leaves a more optimistic taste compared to the previous one. While AR5 hinted at the fate that we are committed to global warming ('Most aspects of climate change will persist for many centuries even if emissions of CO_2 are stopped'), AR6 points out that we could make a difference to global temperatures in the coming 20 years or so. This would be possible if the world were to adopt the assumptions that go into low emission scenarios. This leaves us with two obvious interpretations of the difference in AR5 and AR6. Either the climate system is more responsive to our policy interventions: in this case aggressive and quick GHG reductions can make a discernible difference in the near future. Or the likelihood of such policy interventions is now seen as a real possibility while AR5 assumed this to be unrealistic (perhaps there could even be a combination of both).

If the first interpretation is correct, this would presuppose a different understanding of the climate system, which is unlikely. No big changes in the scientific understanding have been claimed by the IPCC. If the second interpretation is correct, this would presuppose that there is a greater readiness in society to implement climate policies, which has led to a more optimistic evaluation. In fact, there are signs of this, most importantly perhaps the adoption of Net-zero commitments by several governments (see https://zerotracker.net/).

But there is a third possible interpretation. Bearing in mind that the SPMs are approved by governments around the world, it is remarkable that they have agreed to the wording above, i.e. that we could make a difference to global temperatures in the coming 20 years or so. In 2021, governments seem to feel bold enough to make such statements. There are several reasons why they may have felt like this. Among them is a growing concern among the younger generation who will be relevant voters in many jurisdictions in the years ahead.

The overall message over time is that things become worse 'out there' in the climate system, and will become much worse if we continue on our path. A constant in the iterations of reports is that the confidence in science has increased. From report to report we are told that the models have improved. Recalling the distinction between aleatoric and epistemic uncertainty, we can't be sure if we have mastered the former or the latter. The IPCC suggests we have better observations about the past and better predictions about what is likely to happen in the future. The reports are signalling increasing confidence about the human causes of the warming trend: the description goes from 'discernible influence,' to 'likely,' to 'very likely,' to 'extremely likely.' These are subjective probabilities. Climate seems to follow the logic of the cloud, not the clock.

At the same time, the glimmer of hope, just at the time before another important climate summit, has been kept alive. Governments have agreed to include statements about decisive action about GHG reductions, and the possibility to limit warming to below 2°C.

For a long time this was the master narrative on climate change: There is an observed rise in average temperatures, caused by humans through their use of fossil fuels. This is a problem and it is getting worse. Temperature rise has already led to severe impacts and will result in much more, irreversible and possibly catastrophic damage. But there is time to prevent the worst if we start ramping up our efforts now. This means developing ambitious policy measures, including behaviour change, in order to limit global warming to 2°C or less. This narrative is present in public discourse, told many times by people involved with the IPCC, the COP process, UN officials, climate scientists, climate activists, politicians, and journalists.

Just after the heatwave of Summer 2018 in Germany, the influential climate scientist Hans-Joachim Schellnhuber gave a press interview in which he referenced the master narrative. He said that science has been 'repeating a very boring narrative over the past decades. Climate change is real, man-made and highly risky, but we can do something about it. And now we have the experience of the extremely hot summer of 2018. These extreme temperatures – especially at night – have made people realise directly that something uncanny is underway. Add to this the drought, the wildfires – and many have started to think about these issues for which we scientists have provided the background narrative over all those years' (Wille 2018, my translation).

Schellnhuber said he was not exaggerating because the 'house is already on fire.' This apocalyptic metaphor gained wide currency in 2019, not least through the activism of Greta Thunberg. In his opening speech at COP25 in December 2019, UN chief Antonio Guterres said, 'Do we really want to be remembered as the generation that buried its head in the sand, that fiddled while the planet burned?'

But the house-on-fire metaphor also suggests that it is too late to prevent a fire. If it is too late anyway, why bother with climate policies that would prevent the worst? In this logic, all efforts would need to be directed into adaptation policies, not mitigation. But this is not how the narrative is built. Adaptation has always been an afterthought in the global climate change narrative (Pielke et al. 2007) and is not at the forefront of attention today (Berrang-Ford et al. 2021).

Coming back to the narrative policy model (Jones and McBeth 2010), let us consider the aspect of heroes, villains, and victims. The IPCC largely refrains from identifying heroes and villains. It mentions (potential) victims, both human and non-human. Other commentators celebrate heroes, often climate scientists, and dismiss villains, especially the fossil fuel industry, but potentially a long list of actors that are responsible for bad practices. Poor countries and future generations are cast as victims. Greta Thunberg, Antonio Guterres, and a host of climate scientists make use of this rhetorical device. This narrative has been around for a long time, and has been promoted by authors like Gelbspan (1997); McCright and Dunlap (2003); Oreskes and Conway (2012), to name a few.

However, the above narrative also contains elements of confusion. Poorer countries have been cast as victims in the past, whereas the rich Northern countries took the role of villain. This was part and parcel of the design of the Kyoto Protocol which operated on the principle of historical responsibility defined through past cumulative GHG emissions. With the rapidly increasing emissions of China, India, and other countries this principle could no longer hold. The bulk of future emissions will come from fast-developing countries. In 2009 in Copenhagen the time was not quite ripe to redesign international climate politics, but in 2015 in Paris it was. As a result, China was cast in the role of villain in Copenhagen when it resisted to sign up to similar commitments as rich countries. A similar fate awaited India when it resisted more ambitious wording in the Glasgow Climate Pact of 2021.

Commentators also want to develop narratives that motivate climate change awareness and help create efficient policies. A narrative that is functional in this regard is perhaps less committed to 'the truth,' but to the consequences of its message. The anticipated positive outcomes justify the selective use, and misuse, of information. Noble lies are not rare, and commentators of many persuasions make use of them. Of course, they may be over-estimating the effectiveness of narratives, and therefore be of little concern. However, they contribute to a climate of suspicion. This is a problem especially for the official narrative because it positions itself as the authoritative source of knowledge, coming from the best scientists of the world. Alarming statements that are based on selected and exaggerated facts are thought to be justified because they will help convince the public, and policy-makers, to 'do the right thing.' This strategy is bound to backfire. We are no longer in Plato's society where the physician is the only one who understands a medicine, and where the rulers could tell myths about the brotherhood of men.

It is useful to recall the discussion on deliberative democracy in Chapter 5 where we encountered the question if policies and strategies should presuppose a common worldview. Critising Habermas for making such strong demands on deliberative practices, Pellizzoni rightly argued that '[i]t is not necessary for the participants' viewpoints to coincide for co-operation to come about: all that is required is that their individual goals dovetail together. This may be the result of strategy' (Pellizzoni 2001:77–78).

Scientists and activists pushing the alarmist narrative, emphasizing the 'science consensus', seem to be an instance of such problematic expectations of belief alignment. In other words, there is a tacit assumption that only if people believe in climate science (or what they have been told about it) will it be possible to solve the problem.

How effective has the IPCC been?

The most visible policy targets mentioned in climate treaties, the 1.5/2°C limits, are not a product of IPCC reports (Keller 2010). The 2°C limit was adopted by the EU commission in 1996, based on interpretations of scientific results. It gave an answer to the question of what that famous UN resolution, the 'stabilization of greenhouse gas concentrations in the atmosphere at a level that would prevent dangerous anthropogenic interference with the climate system' actually meant. As Randalls (2010:601) observes, '[u]nlike the debates in Europe, the IPCC specifically stayed away from endorsing targets for political action.' However, the panel made use of this limit indirectly, by showing how much more negative impacts would occur when crossing the 2°C. This is done through dramatic visualizations and projections such as the 'burning embers' diagram and use of metaphors such as 'tipping points' (Leemans and Vellinga 2017; Mahony 2015).

While the 2°C limit was based on several years of debate and negotiation, even if it was not 'science-based,' the 1.5°C limit was a political decision which emerged in a short period of time in order to facilitate negotiations and a political compromise in Paris. At stake was the inclusion of vulnerable developing countries which saw the 2°C target as a Northern project that would not prevent most damage to the global poor. At the same time, requests for financial compensations were made by these countries which could be averted by adopting the 1.5°C target (Guillemot 2017:47).

How effective has the IPCC been on climate policy more generally? Here is a pessimistic assessment: 'The direct political effects of the scientific message on the measures adopted have been modest. Scientists have called for heavy reductions in emissions ever since the early 1990s, but global emissions continue to increase strongly. The indirect effect of the IPCC's work may be seen as more significant, particularly in terms of creating increased understanding of this complex issue.' (Andresen and Skodvin 2011:172–173). More and longer reports have been published, and the science consensus has allegedly increased over time (Cook et al. 2013; for a critique, see Pearce et al. 2017). However, this has not had an appreciable effect on political progress, although it has been a constant reference point in public discourse.

I have noted above that the IPCC in AR6 feels confident to make quite alarming statements, and to keep the hope alive that timely and effective policy interventions are possible. My speculation is that the growing concern among the younger generation could explain this shift. The basis for my speculation is that the apparent disconnect between science and politics, as noted by Andresen and Skodvin cannot not explain why climate science all of a sudden became more influential. After all, the IPCC is linked to social and cultural norms and expectations via the participating governments. The same governments meet at climate summits where they negotiate treaties. While the IPCC report is not the formal input into the negotiations, the social forces that have informed it are represented at the negotiations, too. If this is correct, the progress in climate diplomacy and politics is only loosely coupled to climate science and much closer to social and political developments.

Conclusion

There have been many studies critically examining the practices of the IPCC. They have focused on the Climategate scandal, lack of transparency, inaccuracies in the Fourth Assessment Report and the shortcomings of its leadership (see, e.g. Beck 2012; Grundmann 2012). However, there are additional criticisms. I want to emphasize three points.

First, the size and growth of the literature to be assessed by the IPCC poses a problem which is not being faced adequately (Minx et al. 2017). The IPCC is unlikely to fulfil one of its main tasks properly, the assessment of the scientific literature. Leaving the selection of relevant studies to expert reviewers will inevitably expose them to the charge of bias and cherry picking. This seems to be a problem, especially with regard to WG3. IPCC scientists have some power to create, select and communicate data, narratives, and scenarios, thus focusing the debate on specific issues and solutions. This activity is not performed transparently.

As Minx et al. (2017:255) point out, 'there is a crucial role for social science research to assess the "real-world" practicalities of climate solutions. E.g., for building a compelling knowledge map for climate policy alternatives, it is necessary to understand which climate policies do and do not work under specific institutional arrangements, identify the conditions under which individuals or collectives cooperate, or establish how humans behave when faced with different types of pressures or incentives ... However, when it comes to the bulk of the IPCC-reviewed social science literature on climate solutions, progress in accumulating knowledge has been very limited.'

Second, the public discourse is dominated by statements from WG1 about the physical aspects of a changing climate in the future. Results from computer models and sometimes dramatic messages dominate in the media. This cannot be regarded as policy advice but as informed opinion or comment, masquerading as world-class expertise. It serves as agenda-setter, and frames the problem in a specific way, focusing the attention on selected physical parameters, drawing a very unique, one might say: impoverished map of climate change.

Third, the institutional setup of the IPCC prevents it from providing response strategies; it feeds knowledge into the UNFCCC process, via one of its subsidiary bodies, SBSTA. At the same time, commentators close to the IPCC communicate policy alternatives which occupy public attention, frequently based on IPCC scenarios and assessments.

Taken together, the genesis and structure of the IPCC prevents it from fulfilling the function of providing practical expertise about climate change and the options to deal with it. If taken seriously, this task would mean to give priority to the development of policy options and response strategies. In reality, this area of its work is backgrounded by the way WG3 works and is positioned.

Climate change becomes an abstract global thing 'out there,' making us forget that it is related to local, regional, and national circumstances and cultures (Hulme 2010). Climate drivers are located in cities and villages, in factories and farms, propelled by citizens and businesses, but the framing of the problem does not reflect this fact.

As soon as hints towards practical policy options are aired, these do not normally survive the stage where the Summary for Policymakers, and the Synthesis Report are drafted. Nations that see a threat to their interests in some wording will fight hard so that it does not appear in the final report. This restricts the role of the IPCC considerably: it retreats to the now well-known communication of potentially dramatic climate impacts, which the media thankfully disseminates.[7] This recruits a bandwagon of climate activists who think that alarming messages will create the political will, which is seen as necessary (if not sufficient condition) to solve the problem.

The IPCC has not elicited knowledge about the comparative politics of climate change, nor about the question of what works and what doesn't. The self-reinforcing loop between disciplinary activity (witnessed in climate science and climate economics) and IPCC reports thus far has not not extended to social and political science, nor to the humanities.

The IPCC has produced some shared facts that provide reference points for climate debates and policy, world-wide. These are the about the level of observed warming and its causes, and the size of efforts needed to reduce climate drivers to avoid future climate risks. It has done this through a combination of expert roles, of the science arbiter, the stealth advocate, and the commentator. The role of providing scientific advice to policy-makers in the usual meaning is not evident.

The fact that public discourses contain many commentators makes the IPCC one among several voices, potentially eroding its epistemic authority. Some critics demand more reassuring, others more dramatic narratives. They are testament to the fact that the evaluation of the scientific literature is contested, and that the IPCC provides a fragile consensus. Being one commentator among many means that epistemic authority is not a given, but needs to be asserted. As long as the intergovernmental structure is maintained, this task is made easier. Governments like to refer to reports that have been written with their own input. But the demand for independent assessments will never go away, and the unique character of the IPCC might not last.

Notes

1 It is worth noting that resolution 43/53 uses common diplomatic treaty language, such as 'recognizes,' 'confirms,' 'considers,' etc. The resolution mentions ten points and culminates in the verb 'requests,' which is the strongest wording in the list. It reads as follows: 'Requests the Secretary-General of the World Meteorological Organization and the Executive Director of the United Nations Environment Programme, through the

Intergovernmental Panel on Climate Change, immediately to initiate action leading, as soon as possible, to a comprehensive review and recommendations with respect to: (a) The state of knowledge of the science of climate and climatic change; (b) Programmes and studies on the social and economic impact of climate change, including global warming; (c) Possible response strategies to delay, limit or mitigate the impact of adverse climate change; (d) The identification and possible strengthening of relevant existing international legal instruments having a bearing on climate; (e) Elements for inclusion in a possible future international convention on climate.'

Item (c) was tacitly withdrawn in 1992 through the redefinition of IPCC's remit, but as far as I can see the UN resolution was not altered.

2 At the 15th meeting of the IPCC in 1999, Procedures for the preparation, review, acceptance, adoption, approval, and publication of IPCC reports were agreed. Here the term 'policy-relevant' appears for the first time. At the 20th meeting in 2003, the formula 'policy-relevant but policy-neutral' was introduced to describe the nature of the SPM and to describe the questions that could be posed in the Synthesis Report (IPCC 2003).

3 The recent 'Global Stocktake' by the UNFCCC underlines this restricted role of the IPCC, see Milkoreit and Haapala (2017).

4 This has similarities to Entman's definition of framing but is much more comprehensive and dynamic, adding actors and a temporal dimension.

5 Ever since the Third Assessment Report was published in 2001, the IPCC has used the adjective 'unprecedented' in order to make a case for action. The word is used for observed temperature rise, climate drivers (e.g. GHG emissions), technological challenges, and challenges to international cooperation. See Zorita (2019) for a discussion of the flaws of this term when applied to physical parameters of climate.

6 Commentators have identified three main areas of uncertainty: climate sensitivity, future emissions, and carbon cycle feedback (Evans et al. 2021). Carbon cycle feedbacks include the release of methane from thawing permafrost or shifts in forest ecosystems as a result of higher CO_2 levels and higher warming.

7 These are not new. Temperature and sea level rise had been projected in similar orders of magnitude, and discussed in public since the 1980s (Hart and Victor 1993; Grundmann 2015).

Chapter 8

Covid, Expertise, and Society

Stepping Out of the Shadow of Epidemiology?

As in the previous chapter the role of experts and their narratives needs to be emphasized. Similar to the climate discourse heroes, villains, and victims are identified. 'Following the science' has become a slogan adopted by governments and social science observers alike. Success in fighting the epidemic was closely linked to a belief in the pronouncements and models of epidemiologists, the heroes in the story. Those who were not singing from this hymn sheet were seen with suspicion, and those who intonated a different tune risked being cast in the role of the villain, bearing responsibility for the dreadful health effects on vulnerable groups, the victims.

In many countries around the world, governments have set up advisory bodies to develop response strategies against the spread of the Covid-19 virus. Their advice was sought as the virus was new and little was known. The advice has been used to justify policy interventions supposed to steer their country out of the crisis. These bodies are usually composed of scientists from a relatively narrow disciplinary range. Biomedical sciences, in particular virology and epidemiology, are the leading disciplines. Epidemiological modelling provides the epistemic core, the logic, and the rationale for decision-making. The modelling is based on insights from virology, and assumptions about people's behaviour. This privileged position is seen as proper and adequate by most commentators, given the nature of the problem. As a result, experts from the field of epidemiology have provided a narrative for public consumption, and for the justification of government policies. Again, we see attempts at boundary work. The mechanisms of monopolization, expansion, expulsion, and protection are evident but it is not clear if the issue can be kept under the control of epidemiology. Based on a news analysis Erikainen and Stewart (2020) report about credibility contests around DIY face masks and hand sanitisers; around DIY coronavirus treatments and tests; and the delineation of professional science from 'fringe' citizen science. Their analysis shows that the media performed the boundary work, referring to official experts:

> Scientific and public health "experts" were generally attributed with epistemic authority and credibility, even while their claims lacked consensus.

DOI: 10.4324/9781003320227-11

> Relatedly, via discourses around DIY treatments and tests, news outlets engaged in boundary work to differentiate credible "proper" medicine from fraudulent "alternative" medicine… news discourses worked also to differentiate DIY citizen science and maker communities from professional scientists and professional science. They generally rendered the latter safe and effective, and the former a risky even if potentially useful "fringe science" sphere.
>
> (Erikainen and Stewart 2020:10)

Social scientists use similar boundary drawing devices when talking about the Covid-19 'infodemic,' misinformation, and conspiracy theories. Some are searching for the equivalent of a social vaccine, with the aim to inoculate people against wrong beliefs through behavioural interventions (van der Linden et al. 2020).

Epidemiological modelling provides a tool for decision-makers and therefore important elements of practical knowledge. However, this knowledge is incomplete. It needs to identify levers for action in order to become effective, and here we encounter a variety of different possible interventions. They comprise pharmaceutical interventions (such as drugs, vaccines) as well as non-pharmaceutical interventions (NPIs, such as social distancing, mask wearing, restrictions on movement, closure of workplaces and schools, etc.). The scientific expertise of virologists and epidemiologists resides in the first type of intervention, not the second. Nevertheless, it is this type of expertise which is presented as justification for the second type of intervention, which falls in the domain of social science disciplines. However, these are largely absent in the policy advisory systems of most countries.

Giving prominence to the biomedical sciences had significant consequences in terms of framing the problem. Because the problem had already been framed in these terms, we do not question the central role of the biomedical sciences, especially epidemiology itself. Even the appointment of experts for advisory bodies is not seen as a choice, where another choice could have led to other courses of action. It is seen as a 'natural' reflection of the problem.

This is not to say that epidemiology is not important. Of course it is, and it was most instrumental to alert politicians to the exponential nature of the pandemic at specific points in time. Virology provided the important insight that significant non-symptomatic transmission occurs which makes health interventions particularly difficult. Governments felt that it was their responsibility to protect their citizens. Exactly how to do this, and how to measure success, became a matter of debate. Epidemiological models lent themselves especially well to policy interventions as they allowed to link metrics to specific decisions. For example, levels of infections were seen as a matter of concern from which restrictions on public life followed.

We need to recognize the consequences that arise from the privileged position of the biomedical sciences in the process of producing and delivering expertise for political decision-making, and what other perspectives could be useful that have been pushed into the background.

Not only has a particular academic field been privileged, but academic research as such. Yet, expertise is far broader than science based. As I aim to show, there are important sources of expertise in civil society, and in the professions. These sources could contribute more vigorously to a debate about how we should deal with the pandemic, what principles should guide us, and what society we want to build. However, these have been largely sidelined.

Hulme et al. (2020) cogently argue that 'because decisions with far-reaching consequences are being made now, it is precisely the right time to call for a greater plurality of knowledge. This entails bringing in broader sources of knowledge to the decision-making process, promoting more transparent decision-making processes and dismantling unhelpful "hierarchies of knowledge". In these "hierarchies," certain forms of knowledge (e.g., certain disciplines within natural sciences) are seen as inherently superior, rather than as complimentary. Such a diversification of knowledge would benefit both the effectiveness of decisions made, as well as the legitimacy of those decisions among publics' (Hulme et al. 2020).

The necessary but insufficient role of epidemiology

The dominant, epidemiology-based approach has three salient elements (which were developing over time, with overlaps). First, it emphasized the crucial role of diffusion of infections, measured by the reproduction rate R, and the need to reduce them via NPIs. Social distancing, hand hygiene, mask wearing, and isolation were the tools for the job. Then pharmaceutical treatment options were discussed, and recommendations about effective medicines were made. However, an influential narrative developed by epidemiologists was to eradicate the disease, initially through NPIs, later through effective vaccines. This was hoped to be the game changer, allowing us all to return to social life as we knew it. Research and investment in effective treatment has paled in comparison.

Surprisingly, vaccines became quickly available, if extremely uneven across the globe. But they did not hold their promise, and it is an open question if they ever can. To be sure, all of the above measures have contributed to a significant amelioration, but the trend of infections, hospitalizations, and deaths has not been uniformly downward. Waves of new infections were recurring, even in countries with high rates of vaccination. This indicates the limits of the epistemic model which provided the foundation for these strategies and hopes.

The narrative based on the epidemiological paradigm has shifted the goal posts, it has effectively inverted the relation of means and ends. At the beginning of 2021, vaccines were thought to give us the means to overcome the pandemic; at the end of the same year the increase in the ratio of vaccinations has become the end. At least in several countries of the Global North the narrative has shifted from 'We will manage, perhaps eradicate Covid through vaccinations' to 'We need to find ways of greater vaccine uptake.'

It is no surprise that the epidemiological model has come under pressure from its nemesis, vaccine critical voices (Goldenberg 2021). They represent different social groups, and it is tempting but futile to dismiss them as irrational and irresponsible. To be sure, higher vaccination rates lead to fewer infections, severe cases, and deaths. But based on current evidence, the vaccines will not lead to an eradication of Covid-19. There are several reasons for this. The virus has become endemic, and it is present in several animal species. New variants are emerging against which existing vaccines may be much less effective.

The virus, and its ever-evolving mutations, is present in all countries and can move from a high incidence country to a low incidence country. Medical researchers from the USA put it this way: 'Rather than die out, the virus will likely ping-pong back and forth across the globe for years to come. Some of yesterday's success stories are now vulnerable to serious outbreaks. Many of these are places that kept the pandemic at bay through tight border controls and excellent testing, tracing, and isolation but have been unable to acquire good vaccines. … But even countries that have vaccinated large proportions of their populations will be vulnerable to outbreaks caused by certain variants. That is what appears to have happened in several hot spots in Chile, Mongolia, the Seychelles, and the United Kingdom' (Brilliant et al. 2021).

The authors make the perhaps provocative statement: 'The virus is here to stay. The question is, What do we need to do to ensure that we are, too?' Their answer is informed by the epidemiological model, too. The main recommendation is to vaccinate in, and around hotspots of infections, something practiced in the 1970s smallpox outbreak in African countries and India. Such 'ring vaccinations' would solve the problem of shortages of vaccine supply. The authors acknowledge the challenge to their approach but remain optimistic: 'Of course, it was a different disease, a different vaccine, and a different time.'

They continue: 'Part of what makes Covid-19 so difficult to combat is that it is an airborne illness with so much asymptomatic transmission. Today, however, epidemiologists have the added benefit of powerful new tools for detecting outbreaks and developing vaccines. They can use these innovations to build a twentyfirst-century version of surveillance and containment for the battle against this pandemic.'

Asymptomatic transmission of Covid-19 is what makes the virus so persistent. While this was known early on, the fact that it was also airborne came as a later insight. Both elements combine to make existing models of prevention so fragile. This would also be the case with ring-vaccinations. Even vaccinated people can become infected and transmit the virus without showing symptoms. As we have also learned, electronic surveillance systems have not delivered what they were hoped to do.

While the above-mentioned approach recognizes the challenge, and is cognizant of many social variables that are important to consider (including the need to build effective international institutions to deal with this, and other pandemics), it still is heavily influenced by the epidemiological paradigm.

The role of science, expertise, and decision-making

Above I have pointed out that scientific advisors in the pandemic tend to be epidemiologists. Who is missing from the debate? I mentioned the absence of the social sciences, which could provide an academic form of expertise. We also need to consider the contribution from non-academic sources, especially professionals and citizen experts. They bridge the gap between knowledge and decision-making.

Professionals have specialist knowledge of relevant technical aspects or social practices, and the scope for intervention. Lay experts offer views from civil society based on experience, and reflecting social concerns. It should be noted that the World Health Organization (WHO) advocates the inclusion of communities; they should have a voice, be informed and engaged, and participate (Habersaat et al. 2020; Marston et al. 2020; WHO 2020). This lays out the range of expertise which is not always, or has not always been at the table.

Governments have pursued different policies in their fight against Covid-19 (Stevens 2020). Where we see relative success, this does not seem to be due to better scientific understanding, or better communication of the scientific findings, but due to other factors. Some have mentioned preparedness in public health administrations, especially via functioning test and trace and systems, vaccination programmes, adherence to distancing rules, and hospital treatment. Others claim that there is scant evidence for a close link between health infrastructures, wealth, and government regulations on the one hand and success in the fight against Covid-19 on the other. Instead, demographic variables are said to explain most of the variance, especially the age profile of nation states, while other factors usually assumed to be important are less so. One study claims that the Human Development Index (HDI) and the Global Health Security Index are negatively correlated with the probability of dying from Covid-19.

What is more, social inequality as measured by the Gini coefficient, does not predict reported cases and deaths as one would presume. 'These results show that improved indicators regarding economy and capability to prevent and mitigate epidemics and pandemics do not guarantee success in managing the Covid-19 pandemic' (Kumru et al. 2021). In other words, in terms of mortality the Covid-world is upside down to what one would have expected. The authors find that 'Covid-19 cases and mortality have risen as the HDI value, percentage of male smokers, and the elderly population increased, and have decreased as the number of tests per case decreased.'

Covid-19 does not only pose a challenge to human visions of controlling nature. Nature in the form of Covid-19 seems to insult our attempts at understanding. In January 2021, a commentator in the *Financial Times* said, 'A few countries, it is true, have almost unambiguously good stories to tell. But it is some feat to spot the values and institutions that link Israel, the United Arab Emirates and Mongolia. As if to tease us, when a theme does emerge — the heedless "neoliberalism" of Anglo-America — it wilts on further evidence. That is, we have learnt less and less over time. The past 18 months are so haunting in part because they lack all pattern and meaning' (Ganesh 2021). Further pieces of the puzzle are added as we move on. But we don't know how to put them together, or how big the puzzle will be.

The dominance of epidemiology has led to a focus on eradication of the Covid-19 virus. This seemed plausible with the unexpectedly fast development of vaccines. A technical solution seemed in sight which would allow societies to return to normality in the course of months. However, the vaccines have their own problems; they are not 100% safe, nor 100% effective. As with all vaccines they come with side-effects. They also lose effectiveness after months, and are perhaps not effective against mutations of the virus. Pharmaceutical companies are going to benefit from booster vaccinations, maybe for years to come (Oxfam International 2021). At the same time, they reject all liability for side-effects, pushing it onto the purchasing countries, as leaked documents show (The News 24 2021). Their rent-seeking strategy has been rewarded with vast sums of government money, which overall reduces the appetite to spend on other forms of treatment.

Such news gives further support to vaccine critical groups, which in turn poses a serious problem to the epidemiological paradigm, as a certain percentage of immunity needs to be achieved in order for the epidemic to recede. Statisticians Spiegelhalter and Masters reckon this to be around 86% of the population (Spiegelhalter and Masters 2021).

Some authors argue that the problem is lack of information, and the pernicious effect of disinformation campaigns. For example, Baldwin and Lenton (2020) emphasize the eminent role of scientific consensus v disinformation. Like in the climate debate, attempts have been made to emphasize 'the scientific consensus' (Alwan et al. 2020), with some commentators interpreting the Covid-19 case as a problem of misinformation. Extending the narrative on climate change, where 'merchants of doubt' are the culprits for stalling progress, a similar narrative denounces the 'flood of fake news' which allegedly makes people blind to the challenge. This is the well-known tune developed by Oreskes and others. It aligns with the theme of the 'villain' in the policy narrative presented previously, and of course with the boundary work mentioned above.

Attempts at controlling the information infrastructure, for example, by banning certain voices, or by emphasizing 'correct information,' may have the opposite effect from its intended goal. In both cases belief formation should be understood in terms of 'cultural cognition' (Kahan 2012). As several studies have shown, vaccine hesitancy is a complex phenomenon (Hobson-West 2003; Reicher and Stott 2020). Issues of trust vis-à-vis the government and its public health institutions are crucial in this respect. Information campaigns will change very little, especially if they are 'talking down' to those who are not convinced.

What is more, official experts are on record of telling 'noble lies.' A prominent example is the issue of mask wearing during the early phase of the pandemic. Health experts in several countries have claimed that there is no scientific evidence to show the effectiveness of masks in reducing infections. Perhaps the most prominent case is Anthony Fauci (Cooper 2021), who told this noble lie in order to prevent a run on masks which would impact the supply for doctors and nurses. He has also been called out by the *New York Times* for misrepresenting the Wuhan lab-leak story, the health risks to Hispanic and Black children, and the effectiveness of vaccines (Stephens 2021).

Of course, noble lies are part of political rhetoric. In this regard little has changed since Plato's time. However, there is a difference between politicians trying to manipulate the electorate and experts who claim to represent scientific truth.

National responses

The spread of the virus is uneven across the world. Despite gigantic efforts of individual countries to get their citizens vaccinated, new variants of the virus are likely to appear anywhere, and spread across the globe. There is no global governance institution to coordinate the response measures. The powers of the WHO are limited and its advice to governments has been disregarded in the past. Attempts at international coordination and cooperation are minimal (Brown and Susskind 2020; Buck et al. 2020; Grundmann 2021), although some collaboration has been underway in the search for a vaccine through the COVAX initiative (Kupferschmidt 2020).

National responses have been the standard mode of operating. Vaccine nationalism has been evident in the procurement and distribution of vaccines, and in the politics of recognizing vaccination certificates across borders. Most importantly, rich countries have secured the lion share of vaccines, leaving large parts of the rest of the world without.

Throughout the pandemic a set of important questions was never answered unambiguously: What is our aim? What is the strategy? Sometimes some governments have stated the eradication of the virus as their goal ('zero-Covid'). At other times more modest goals have been announced: contain the spread of the disease, reduce hospitalizations and severe cases, keep the health system going. Some have argued that 'living with Covid' is the only realistic option (Brilliant et al. 2021). This involves several measures, like continuing with social distancing, mask wearing, and providing effective treatment in severe cases. In this perspective, disrupting social life is not legitimate, especially when the number of severe cases and deaths reaches low levels comparable to other diseases.

Wicked problems

Wicked problems only can be managed better or worse, not be solved once and for all (Grundmann 2016; Rayner 2006; Rittel and Webber 1973). The verdict is still out if Covid-19 falls into this category of social problem. The quick development of vaccines has given hope that successful vaccines will solve the problem. Initial enthusiasm has given way to a more sober assessment as the virus has shown several mutations which may hinder the vaccine effectiveness (Vogel and Kupferschmidt 2021).

Nevertheless, the availability of vaccines could be seen as a 'technological core' which can be refined over time (Sarewitz and Nelson 2008) so that herd immunity can be achieved. Cheap vaccines that can be easily stored and administered would help. However, as the above has shown there is not always sufficient supply of vaccines across the world, new variants are likely to emerge, and vaccine hesitancy may be a force that could limit such efforts. Wicked problems have no stopping rules; if zero Covid is not a realistic prospect we will see different definitions of success come and go. Political posturing, and creative use of statistics to prove success is to be expected.

Given these parameters it appears that the involvement of all forms of expertise would be beneficial for developing effective policies. A broader range of expertise apart from biomedical scientists would help. As people became frustrated by the on-off logic of large lockdowns that are based on territories and industries, other solutions were considered.

What are the prospects for expertise from other parts in society? Engineers may develop standards for well-ventilated rooms in which it is safe to congregate for prolonged periods of time. Community leaders may develop early warning systems that allow for timely interventions. Policy-makers could try to find solutions to the competing jurisdictions between private and public bodies, and between different levels of government. They also could incentivise efforts to find effective treatments, and develop strategies in the first place, taking the initiative away from private corporations who are the 'big players' in this saga.

Health professionals may develop testing regimes that are quick and reliable and allow targeted responses. Global institutions need to ensure that resources to fight the pandemic are distributed fairly among countries. Pressure groups could campaign against the rent-seeking behaviour of big pharma. In sum, the governance of Covid-19 is something that needs to be established as a problem. Only when it is recognized as such will we be able to get away from an important but overly narrow definition of the problem, and open new avenues for intervention.

To sum up, I want to emphasize three points. First, the dominant approach to deal with the pandemic is based on a biomedical framework. While this provides important insights and tools for intervention, it is far too limited to address the manifold problems that the pandemic poses.

Second, it is not enough to extend the range of expertise which is considered in decision-making beyond biomedical research; it also needs to be extended beyond science. Professional bodies and civil society are not part of advisory bodies and thus not represented in decision-making. Drawing on the insights from PNS, lay expertise and deliberative democracy, concerns and interests need to be articulated and mediated, resulting in courses of action that are broadly supported.

Third, hopes about a quick end to the pandemic via vaccinations have been dashed. The virus is endemic and bouncing back from many corners of the globe. Both the pandemic and the vaccine deployment are unevenly distributed. Effective policy responses require long-term cooperation of many societal actors. The governance of the pandemic, and the role of expertise, broadly conceived, is the problem that needs addressing.

Covid and climate: Similarities

Several studies have compared Covid-19 to climate change, emphasizing specific aspects. Buck et al. (2020) look at Covid-19 and solar engineering, highlighting the role of policy goals and metrics, models and scenarios, different time horizons, and international cooperation. Baldwin and Lenton (2020) compare ozone depletion, climate change, and Covid-19. They emphasize the need for a science-led policy response and the dangers of disinformation campaigns. Hulme et al. (2020) compare the different meanings of the term 'crisis' in both cases, Lidskog et al. (2020) compare the different spatial and temporal aspects in both cases. I will pick up some issues raised in these contributions, but also refer to other recent research about Covid-19.

In both cases scientific uncertainties persist. In climate science one of the most fundamental uncertainties relates to climate sensitivity, the long-term temperature increase as a result of greenhouse gas (GHG) concentrations, usually gauged as a doubling of pre-industrial concentrations. Scientists have come up with a range, not a definite number. The range is large enough to accommodate vastly different policy options (Grundmann and Rödder 2019). In the case of Covid-19 there are arguably many more unknowns, not least due to the fact that this is a new disease and science needs time to construct an evidence base (Daston 2020; Pearce 2020).

The climate change discourse in many countries is dominated by scientific voices. Scientists act as commentators in public, sometimes in a role of an open, sometimes of a disguised advocate (Pielke Jnr. 2007). Their narrative is that the science is clear, time is running out, and we need to act now. Massive GHG reductions are needed to avoid catastrophic consequences, but the worst can still be avoided. The alleged 'scientific consensus' is the bedrock on which actions should be based.

What is more, the communication of this consensus is seen as an important element in a strategy to bring about ambitious climate policies (Cook et al. 2013; for a critique see Pearce et al. 2017). Those who disagree are easily marginalized and ostracised as deniers, 'merchants of doubt,' or 'anti-science'

(Oreskes and Conway 2012; for a critique see Grundmann 2013). Some influential climate scientists try to establish themselves as authoritative voices in public discourse, and denigrate those who disagree as ignorant, interested, or in other ways illegitimate. They are portrayed as the villains who are ultimately responsible for the lack of decisive climate policies.

Recall the critical remarks from Tetlock above, which could be read as an apt, if somewhat harsh comment:

> On one hand, overconfident experts may be more quotable and attract more media attention. On the other, overconfident experts may also be more likely to seek out the attention. The three principals—authoritative-sounding experts, the ratings-conscious media, and the attentive public—may thus be locked in a symbiotic triangle. It is tempting to say they need each other too much to terminate a relationship merely because it is based on an illusion.
>
> (Tetlock 2005:63)

In the case of Covid-19, a similar narrative of urgent-action-is-needed-to-avoid-the-worst has been developed. The core assumption is that the virus can be eradicated and that we can achieve a zero-Covid world. For a variety of reasons, this may turn out to be wishful thinking. The narrative also fails to accommodate the fact that we have already seen catastrophic results in terms of excess deaths, in several countries. The narrative is still general enough to accommodate these calamities, by emphasizing that 'the worst' could still be avoided.

Each wave of infections, hospitalizations, and deaths can be seen as another stretch, plunging us further downhill. This makes it quite difficult for epidemiologists, and other experts on the topic, to claim the mantle of expert authority, especially if there are unexpected developments that could be seen as going against the official narrative. Controlling the narrative is the ultimate tool for politicians in power. This requires frequent updates and revisions to earlier versions as the pandemic landscape keeps changing.

The Omicron variant which appeared at the end of 2021 posed a challenge to the modelling community. Predictions about hospital admissions seemed to be far too pessimistic. In the UK, the advisory group SAGE predicted 'that Omicron was between three to eight times more transmissible than Delta – but did not appear to make any adjustments in the models for severity. They assumed the virus was as deadly as Delta but by then real-world South African data was clearly showing the variant had far less severe outcomes' (Simmons 2022). This was a surprise to many, and required an update on the narrative. The Omicron wave produced far more infections than previous variants but did not lead to a proportionate rise in hospitalizations and deaths. What is more, many vaccinated people could catch and transmit the virus, even if not becoming severely ill. The diffusion of the Omicron variant happened at a time when several countries were pondering mandatory vaccination.

This gave a boost to vaccine critical voices as it seemed to be questionable to carry on with the idea of regular vaccine boosters. If no effective vaccines can be delivered in time, and the effects of the new variant are milder, then the case for letting the virus spread becomes stronger. Scientists with a track record of being cautious have indicated this course of action might be inevitable, effectively advocating herd immunity through infection (Specia 2022). The changed policy amounts to managing the wave of infections, not to break it, to use a popular metaphor. Boundary work undertaken by epidemiologists became more difficult as their credibility was called into question.

Both climate change and Covid-19 are areas of research with a very fast rhythm of publications. In the previous chapter we have seen the doubling time in climate change compared to the average doubling time in scientific publications. According to the Web of Science (search conducted on 14 September 2021), there have been nearly 320,000 articles published which mention the term 'climate change' or 'global warming' in the topic (title, abstract, or key words). This count covers a time span of over a century, with more than half of these publications coming out in the last five years.

Since 2020 nearly 170,000 articles that mention 'Covid' or 'Sars-CoV2' have been published. This is roughly double the publication activity in the field of climate change, which is already a very fast-growing field. In addition, we see a high volume of preprints in the case of Covid-19, many of which may be relevant for experts advising governments. Some of these papers are hyped in the media before they are peer reviewed. Many never make it into a journal and some do turn out to be wrong and have to be retracted (Rubin et al. 2021).

How do experts keep abreast with the ever-changing research landscape? We know little about how this process is organized. Research on climate change is assessed by the IPCC, even in the absence of systematic reviews, a notable fact (see above, and Minx et al. 2017). There is no equivalent to the IPCC in the case of Covid-19. And systematic reviews get published whenever researchers happen to undertake the task. For Covid-19, nearly 180 such publications are on record (as of September 2021). This does not make the task of gaining an informed overview any easier, and one would expect a review of reviews. But this task would need to be organized by a body that performs the role of a science arbiter. Given the time pressures to act on available information, governments across the world organize the process of producing assessments according to their own needs, rules, and priorities. In any case, claims such as those by the UK government that it is 'following the science' are misleading euphemisms. As Cairney (2021) put it, 'Guided by "the science" means "our scientists," and usually a small group of government scientific advisors.'

A final similarity relates to the role of modelling which is evident in both cases. Climate models and scenarios have dominated scientific research and been communicated widely in public (Borie et al. 2021; Grundmann and Stehr 2010). These can be useful tools to get an idea about possible future developments. However, there is always a risk to interpret model results as predictions which are then seen to be falsified by real developments, thus undermining their credibility.

In climate change, integrated assessment models (IAMs) have incorporated physical and economic aspects into ever more complex models (see Peng et al. 2021). IAMs have been criticized heavily, mainly on methodological grounds (see Pindyck 2017). With Covid-19, epidemiological modelling, while dominant, has not included economic, social, political, or psychological dimensions. For better or worse, there are no IAMs in Covid-19. Nevertheless, in both cases models are used to justify policies. The use of best-, and worst-case scenarios is a contested matter. Again, both extremes are used to justify policies. As pointed out several times in this book, expert narratives frighten people, but also reassure them.

To sum up: both cases are similar in that they show scientific uncertainty, pressure to take decisions, alleged science denial, lack of systematic reviews, and a central role of models and scenarios for legitimizing policies. Scientists try to engage in strategies of boundary work, with mixed prospects of success. Expert narratives tell about doom but also give hope; they feature heroes and villains. Experts have made use of noble lies hoping to advance their case. I now turn to the differences. They may be technical but are embedded in a moral that everyone understands.

Climate and Covid-19: Differences

The policy goals in climate change are stated explicitly in international commitments via the process under the United Nations Framework Convention on Climate Change (UNFCCC): stabilization of climate, avoiding dangerous warming, i.e. limiting a warming increase to under 2°C. This has been agreed by the Conference of the Parties in 2015. Nations are supposed to implement this policy through a 'pledge and review' process. McLaren and Markusson (2020) argue that there was a shift in policy goals over time. It started in Kyoto (1997) with percentage emission cuts, followed by a focus on concentrations in Copenhagen (2009), then on cumulative carbon budgets in Doha and Durban (2011, 2012), and finally on temperature targets in Paris (2015). I would argue that while such differences between such goals can be detected, they are closely linked.

The carbon budget is a numerical value that has its equivalent in a number that expresses atmospheric GHG concentrations, and in a number that projects future warming, via the concept and metrics of radiative forcing. In this sense, there has been little change over time: the different metrics are subordinated under the policy goal of avoiding dangerous warming, and they can be translated into each other. I know it is more complicated than that; there are ranges of numbers, and model assumptions that can lead to different outcomes in models and scenarios. What matters for the present purpose is the stability of the main policy goal over time, and a battery of stabilized metrics.

With Covid-19 things are different. There are metrics that translate into each other, but only for a short time period. The number of infected people does not equate over time to a number of hospitalizations, ICU admissions, or deaths. Progress in treatment and prevention have changed these ratios fundamentally. New variants of the virus may change this again.

International cooperation

In 1992, an international governance structure was established by the UN which ever since has institutionalized climate change as a topic for governments, and the legal instruments to deal with this issue through a mechanism of ratification of international treaties (Keohane and Victor 2016).

No such international governance body exists for Covid-19. Attempts at international coordination and cooperation are minimal (Brown and Susskind 2020; Buck et al. 2020), although some collaboration is evident in vaccine development and delivery (Kupferschmidt 2020). The international body which is relevant here, the WHO, has a broad mandate which is 'the attainment by all peoples of the highest possible level of health.' In order to achieve its objective, it sees its function as being the directing and coordinating authority on international health work. It establishes and maintains effective collaboration with the United Nations, specialized agencies, governmental health administrations, and professional groups. It assists Governments, upon request, in strengthening health services and furnishes appropriate technical assistance and, in emergencies, necessary aid upon the request or acceptance of Governments (World Health Organization 2005).

This brief comparison shows how in both cases we see a difference in the kind and scale of response strategies. While the IPCC provides unified metrics, it does not have an advisory function, as I have shown in Chapter 7. Climate change is being addressed through internationally coordinated research assessment (via the IPCC) and broad policy goals (via the UNFCCC), something which is lacking in Covid-19. The WHO does not provide standardized metrics (and it could not possibly do so, given the novelty of the disease and the lack of an established evidence base). But it does have an advisory function across its member states.

The role of data and metrics

Ever since governments started using data in a systematic way to aid decision-making and justify decisions, the interested use of data as a political weapon has been a major concern. This comment on 18th century England captures it well:

> Ministers, parliamentarians, and pamphleteers were all prepared to use numbers to establish or promote policy positions. Such debate was, of

> course, often self-interested and the numbers employed were unreliable if not downright lies–the point is simply to stress the evident willingness which existed to employ quantification as one weapon in the arsenal of the politician, commentator, lobbyist, and projector.
>
> (Hoppit 1996:531)

As outlined above, the IPCC has developed an accepted toolkit of metrics. It has set up national GHG inventories, and agreed-upon methods to measure emissions and emission reductions. This task proves to become especially relevant after the Paris Agreement as countries engage in a pledge and review process where reliable and uncontroversial data is needed. A special Task Force developed and refines an internationally agreed methodology for the calculation and reporting of national GHG emissions and removals, and to encourage the use of this methodology by parties to the UNFCCC (Siebenhüner 2003; UNFCCC 1992).

As a result, to take one example, no country disputes the figures about their domestic GHG emissions. However, while the IPCC is able to frame the problem in specific ways, it is not in a position to provide advice to governments; it does not suggest policy measures to decision-makers (Baldwin and Lenton 2020; Bodansky 1993; Pielke Jnr. 2007; Sundqvist et al. 2018). As indicated, there are still many uncertainties in climate science, the most relevant relate to climate sensitivity, future emissions, and carbon cycle feedbacks.

In the case of Covid-19, relevant metrics include the number of infections and vaccinations, the reproduction rate, number of tests, hospital admissions, or mortality. Countries have different data collection and presentation systems. Even where they use the same software platforms, e.g. the ArcGIS Experience Builder, the range, format, and detail of data is not comparable across countries. In addition, unreliable procedures at the national level make the problem worse (Clough 2020). The authors of one of the platforms that present comparative data have complained that '[t]o properly monitor the impact of the pandemic we would need metrics that are unfortunately not available' (https://ourworldindata.org/coronavirus).

Reliable data is hard to come by, indicating deep uncertainty. Mortality is especially intriguing. Surprising as it may be, we do not know the true death toll caused by Covid-19. An overview article in *Nature* highlights several problems (Adam 2022). Official figures say that by January 2022 5.5 million people had died of the disease. Official numbers can under- or over-report. For example, the Netherlands counted only those who died in hospital after testing positive for the virus, while Belgium included deaths in the community and everyone who died after showing symptoms of the disease, even if they weren't diagnosed (Adam 2022).

Official numbers are not reliable and it is not clear how to arrive at a true estimate. Excess deaths have been used as proxy but this metric has its own limitations if it is not adjusted for age. Based on raw data, excess mortality in Germany was estimated to show 5% more deaths in 2020 compared with 2019. After taking Germany's age demographics into account the number was said to be about 1% (De Nicola et al. 2021).

Various modelling exercises estimate the number of deaths at two to four times as much as the official figure, or between 12 and 22 million excess deaths. *The Economist* has used machine learning technology to estimate the death toll and conjectures that India has probably ten times as many deaths to mourn compared to the official figure of half a million. In China, the true number is 750,000, or 150 times greater than the official number (Adam 2022).

Based on this model, findings across different countries in the world contradict the official view that mortality has been highest in the rich countries. 'Overall, the model suggests that lower middle-income countries (as described by World Bank groupings) have suffered at least as severely in per-capita deaths as rich countries … That's despite the fact that these poorer countries have younger populations' (Adam 2022). However, if it is true that mainly older people die from Covid (Dowd et al. 2020; Levin et al. 2020) it would be expected that countries with a younger age profile would have lower fatality rates. Low-income countries confirm this expectation, no matter which metric is applied.

Overall, there is a sense of frustration among researchers. The above-mentioned article in *Nature* cites a statistician who is leading a modelling project run by the WHO to estimate the pandemic's excess death toll. He thinks that the modelling is intrinsically flawed: 'The data are a real mess and so any modelling effort is going to be very speculative …it's very frustrating as the data are so limited. I'm not happy with the assumptions we're being forced to make, but we're doing the best we can.'

The last problem of uncertainty I want to mention is the long-term effects of the virus, also known as long Covid. There is no universally agreed definition of this condition. It covers a broad range of symptoms such as fatigue, muscle pain, and difficulty with breathing or concentrating. Again, we do not have reliable data on the prevalence of this condition. Estimates put the prevalence at 10% and more of people infected with Covid-19. The UK Office for National Statistics wrote in 2021 that it 'estimates that over a million people in the UK were reporting symptoms associated with long Covid at the beginning of March 2021, with over two-thirds of these individuals having had (or suspecting they had) Covid-19 at least 12 weeks earlier. An estimated 674,000 people reported that their symptoms have negatively impacted on their ability to undertake their day-to-day activities. People who tested positive for Covid-19 are around eight times more likely to suffer prolonged symptoms than observed in the general population' (ONS 2021).

Callard and Perego (2021) report on the emergence of long Covid and the role of patients and citizen experts gaining epistemic credibility, even epistemic authority: 'Long Covid has a strong claim to be the first illness created through patients finding one another on Twitter: it moved from patients, through various media, to formal clinical and policy channels in just a few months. This initial mapping of Long Covid – by two patients with this illness … demonstrates how patients marshalled epistemic authority.' They conclude, 'Patient knowledge needs to be incorporated into how Covid-19 is conceptualised, researched, and treated.' However, this claim has not received recognition through official medical gatekeepers as the definition and mechanism are not established.

The role of norms and values

Where political decisions are justified by scientific information but the science cannot provide certainty or practical solutions, the role of values becomes especially important. No matter how much we know, what we see as a solution to a problem needs to resonate with our deeply held convictions about what is right and wrong (Hulme et al. 2020). These values enter the experts' judgement when developing courses of action. Those who are most vulnerable to climate change impacts tend to be poorer nations, and future generations. In terms of values, solidarity and individualism are conflicting (Verweij et al. 2006). This value conflict manifests itself in the many difficulties and failures to implement effective climate policy goals.

The most vulnerable in the Covid-19 pandemic are the elderly and the socio-economically disadvantaged groups, often in poor health (Horton 2020). The principle of (tacit) solidarity arguably underpinned blanket mobility restrictions, as the young had not to fear so much from the disease but were still included in the restrictions (Buck et al. 2020). Solidarity raises the question of fairness: if the burden from government controls falls disproportionally on different groups, interest mediation and cross subsidies are required. Again, in contrast to Covid-19, the UN has established several mechanisms to address some of these concerns, for example, the Green Climate Fund.

Some prominent voices have made the case for freedom and economic well-being, pointing to dangers of totalitarianism and disastrous effects of Covid-19 restrictions on people's wealth and health (Celermajer and Nassar 2020). Letting the virus spread among the young has been suggested, while still shielding the vulnerable from infection. Utilitarian arguments have been put forward about the trade-offs of continuing Covid-19 controls, and their overall negative social costs (see Eyal 2020:22 for critical discussion).

Protecting the vulnerable is a collective action problem. Allowing individuals taking personal risks is not a sensible strategy from a public health perspective, as this would make the spread of infections more likely. Government-imposed controls have thus been enacted, limiting individual liberties. This leads to an interesting difference when it comes to government controls in the case of climate change and the pandemic. Pondering the repeated failure of ambitious climate policies, Roger Pielke Jr. established 'the iron law of climate politics.' It says that 'when the environment and economy are presented as trade-offs, the economy invariably wins' (Pielke Jnr. 2010:49).

From what we have seen so far in the Covid-19 crisis, this law does not apply here. Governments across the world have adopted measures that have severely hurt their economies. They seem to have struck a compromise between these different and opposing values which were embodied in epidemiologists' plea for restrictions on personal freedoms and other social groups wanting to go back to 'normal' patterns of social interaction. The compromise is not stable but temporalized: public life, at least for a while, has assumed a nervous rhythm of opening up and shutting down.

Over time, the value of solidarity in the Covid saga has waned. Here I do not refer to the relation between younger and older generations, or between healthy and vulnerable groups. The issue is transmissibility of the virus through vaccinated people. Before this became widely known, the argument about solidarity was that to get vaccinated meant to help protect the vulnerable. A new situation arose where it was in the best interest of everyone to get vaccinated for personal protection. The fact still remained that the risk of becoming severely ill, or to die, is greatly reduced through the jab. The narrative was adapted to emphasize the personal advantage from being vaccinated.

Crisis and emergency

A final important difference between the two cases relates to the time dimension. Since the 1960s, scientists have thematized climate change as a problem which will occur in the future. It is a long-term process where effects of human interventions are visible after decades to centuries. Climate scientists have warned the public about the problem, and tried to persuade them about the reality of the risk. Politicians by and large have now acknowledged the problem, but postpone difficult and unpopular decisions during their term in office. However, some effects of rising temperatures are manifest today and have prompted some protagonists to use the rhetoric of crisis and emergency.

The dynamics of Covid-19 is different. It plays out in days, weeks, and months. This difference in time horizons has led to a different sense of urgency, and to a focus on the level of decision-making which is quick and effective (Buck et al. 2020). Large parts of the publics in many countries do not need to be convinced that there is a need for action. Governments imposed restrictions that would have been difficult to imagine a few years ago. They are conducting large-scale experiments with their populations, on their territories (Elden 2007). This raises many questions, among them the proper acknowledgement of multiple forms of knowledge and normative criteria (Hulme et al. 2020), or the question of where the most effective levers for intervention can be found, and if the national focus neglects the local, regional, and international dimension. Given the short-term logic of the problem, consequences of policies become manifest during one term of office. National governments thus have the incentive to perform well, or play the blame game, and find scapegoats (Pearse 2020).

Conclusion

Despite some similarities between both cases, what stands out are the many significant differences. While both cases show limits in scientific knowledge, similarly structured policy narratives, and the important role of norms and values in decision-making, there are differences with regard to reliable data, policy goals, international cooperation, and time-horizons. In the climate case standardized data is available, which is linked to clearly defined policy goals. With Covid-19 such metrics do not exist, and the policy goals are shifting all the time. In both cases there is a focus on national dimensions of policy-making, but climate change also has global governance institutions which have established agreed metrics and policy goals. With Covid such goals are manifold, different across countries, and shifting over time. Climate is a long-term problem that has been postponed by governments, while Covid-19 is happening in real time, forcing politicians' decisions.

Different policies are developed on the basis of value commitments and appeals to public support. I have focused on freedom and solidarity as partly opposing values, linking freedom to a logic of individualism and libertarianism, and solidarity to a concern about future generations, vulnerable people, and a public health ethos. Through international institutions, climate change has established a more stable compromise between these, whereas Covid-19 still is in a fragile limbo. Here, government restrictions have led to unprecedented economic impacts. Such impacts were to be expected (stay-at-home orders lead to a slowdown of economic activity). It was therefore difficult to predict such interventions, given the prime goal of economic growth in most countries.

Does the comparison enable us to draw lessons? My focus has been on an analytical level and I do not think that the issues identified here could be changed at will, within short time periods. Compared to Covid-19 climate change seems to be a slow-moving issue. It has established international institutions and data infrastructures. Covid-19 is fast moving, and the building of international institutions has not been a priority, perhaps in the belief that the problem will be over soon. This belief may turn out to be wrong. The hard work of developing international cooperation would seem prudent.

What do climate change and Covid-19 tell us about the expertise which is deployed? In both cases there is a privileging of scientific voices that are rooted in modelling. This privileged position reflects the hierarchy of knowledge in society, and the social status that comes with it. Activists, and scientist-activists, find it natural to use worst-case scenarios to initiate and legitimate government policies that impose restrictions on certain types of social practices. Climatologists and epidemiologists have had free reign to frame these problems in particular ways. Wider audiences have become accustomed to follow their lead, and to accept their problem definitions as natural. The same applies to the solutions, insofar as these have been proposed.

This allows scientists to play an influential role as public commentators, and as advocates, while officially being scientists who only report on 'the facts.'

Other forms of expertise have been regarded as less legitimate as their contribution to solving the problem of global climate change, and the problem of a pandemic is allegedly beyond the grasp of civil society and the professions. With this came a tendency to define the problems, and their solutions, in scientific and technical terms.

A final point on narratives. As many policy-related narratives have the elements of framing, plot, drama, and moral we tend to encounter the protagonists in the roles of heroes and villains. These roles allow to simplify complex stories, to state causality, and to attribute blame. Al Gore, the IPCC, and Greta Thunberg are the heroes in the fight against climate change. In a similar way, epidemiologists are the heroes in fighting the Covid-19 pandemic. Villains in both cases are those who debate the science, or reject far-reaching policies that would disturb their business or lifestyle. The heroes attribute the blame for a lack in progress on the villains, who are demanding unreasonable standards of proof, are plain anti-science, or irrational. This simple narrative is present in the Hey Jude theme played by Naomi Oreskes and others, and resonates with the belief system of many left-liberals of the Northern hemisphere. There is no challenger discourse which has the same simple structure, but many critical voices: commentators are everywhere. The science base, and policy decisions, can be discussed by anyone. The Star-Spangled banner, Hendrix version, is difficult to be imagined as a sing-along smash hit.

Chapter 9

The Challenge to Professional Expertise

In this chapter, I will present three case studies of professional fields where official expertise has been challenged by so-called 'lay experts' and technical expert systems, known as artificial intelligence (AI). The cases are medicine, journalism and law. I will present and discuss some salient contributions, focusing on the different types of expertise and their narratives.

Contrary to much of the public debate, which is focused on issues of job losses due to the introduction of AI systems (Autor 2015; Frey and Osborne 2013), my interest is to explore the question of how such systems will change the configuration of expertise in society. Many social interactions have been mediated by technological devices for a long time. Some authors claim that the massive introduction of AI systems will provide a step change, be pervasive, and transform societies in radical ways. However, commentators disagree about the extent and desirability of this change.

Narratives play a role in this, too. Except they are not linked primarily to a policy issue (climate policy, Covid-19 policy) but to a more general account of the world of the professions, what it is, what it will be, and what it should be. Professional experts and commentators are involved in creating and disseminating narratives. They espouse their views, and representatives of various interest groups are keenly involved. The issues are framed accordingly, and we do encounter archetypical patterns of plot, heroes and villains, and morals. While the rise of citizen expertise is often seen in a favourable light, the evaluation of its impact on medicine, law, and journalism is not uniform. Thus a prominent narrative attributes the rise of Post-Truth to the rise of non-expert voices in public discourse. The advance of AI is seen as even more controversial. For some it is unquestionably a good thing; for others it is likely to lead into a dystopian future.

In their book *The Future of the Professions*, Richard and Daniel Susskind convey a positive view. They argue that the major problem of expertise today is its unequal distribution. While the expertise is there, so to speak, it is not available to most. The solution is to spread expertise at lower cost. But the professions as they are currently organized stand in the way. They have a vested interest in keeping their expertise tight and overcharging for its use. While new technologies have

DOI: 10.4324/9781003320227-12

provided the preconditions for a more equal distribution of knowledge in society, antiquated monopolies have blocked progress. The Susskinds are playing the tune composed by Koppl, Berger and Luckman, and others.

While it is undisputed that To be sure, professional organizations act out of self-interest, but this is not the whole picture. It is problematic to adopt a narrow aim of increasing overall societal efficiency and economic productivity, without considering the interests and values of different social groups. It also shows a narrow notion of expert knowledge and expert activity. The authors treat expertise as a given, as something that delivers certain knowledge and advice.

As a result, they do not see the task of experts making judgements in the face of uncertainty or conflicting values. Experts are simply collecting and interpreting information, a task that can be transferred to intelligent machines. Large parts of their book are dealing with possibilities of this transformation or delegation process. The transformation is conceived in either of two ways, automation, and innovation. As they put it, '[w]hereas automation is the use of technology to support this traditional model [of expertise, R.G.], innovation enables ways of making practical expertise available that simply were not possible (or even imaginable) without the system in question.' (Susskind and Susskind 2016:112).

Machine learning is the technical term which sums up the novel approach to data analysis and prediction. A common definition is 'the use and development of computer systems that are able to learn and adapt without following explicit instructions, by using algorithms and statistical models to analyse and draw inferences from patterns in data.' We are all familiar with examples such as speech recognition or face recognition.

It involves feeding the system with example inputs of the task that needs to be accomplished. The system is trained by data input which enables it to learn. The algorithm makes its own decision regarding the operation to be performed to accomplish the task in question. This approach is different to previous computer-based systems, such as statistical modelling or brute force algorithms.

Traditional statistical modelling seeks to simulate real-world processes by using data. One can talk about predictions only to the extent that the model output fits well with the real-world processes it is supposed to simulate. This means a programmer needs first to specify a mathematical equation expressing an outcome variable as a function of selected explanatory variables put together in a particular way, and then to see how well the data fit with the analyst's choices. Regression analysis is one such technique, and it provides information about the magnitude and direction of relationships between variables that are selected and specified in advance.

This is not the case with machine learning, where the algorithms allow the data itself to dictate how information contained in input variables is put together to forecast the value of an output variable. Coglianese and Lehr (2017:1158) pertinently ask: 'Given the existence of these alternative statistical techniques, what advantages do machine-learning algorithms offer? Its proponents claim that they outperform standard procedures in terms of predictive accuracy and statistical efficiency (that is, the increased ability to obtain predictions with both low bias and low variance)'.

In 2016, the computer programme AlphaGo, perhaps the best-known machine learning device in recent years, made the headlines when it beat the champion in the board game Go. Unlike previous success of computers against humans in the game of chess, this was not based on a brute force approach where all possible moves are calculated at any given point.

Not everyone shares the enthusiasm about AI systems. Many see them with suspicion. They are said to be intransparent even to their creators (Pasquale 2015). The critique is based on the argument that we should not rely on decisions which we cannot rationally follow or reconstruct. For many reasons opaque decision processes are deeply problematic. Burrell (2016:1) offers a distinction between three forms of opacity: '(1) opacity as intentional corporate or state secrecy, (2) opacity as technical illiteracy, and (3) an opacity that arises from the characteristics of machine learning algorithms and the scale required to apply them usefully.'

It is important to identify 'the form (or forms) of opacity entailed in a particular algorithmic application in order to pursue a course of action that is likely to mitigate its problems' (Burrell 2016:3).

Predicting a brave new world of automated expertise, Susskind and Susskind are unfazed by such issues. They declare: '[J]ust as we are coming to terms with the idea of driverless cars, so too we will feel comfortable with the concept of teacherless students, doctorless patients, lawyerless clients, consultantless businesses, clergyless parishoners, and so forth' (Susskind and Susskind 2016:42).

Is this hype or reality? Consider the case of education where professional power has been challenged through the development of new technologies, against the backdrop of de-professionalization of colleges and universities. Blended learning, online courses, and more sophisticated learning analytics are examples given by the Susskinds. They predict that these developments will change university classroom teaching from a model described as 'the sage on the stage' to a model called 'the guide on the side.' But this has not become reality yet, not even during the Covid pandemic where teaching has been conducted online. Despite the largescale adoption of new learning technologies the predicted transition has barely materialized.

Susskind and Susskind admit that small group tutorials, as practiced at Oxford and Cambridge, are remarkably effective. The challenge, as they see it, is to emulate the model through an online system that obviates the necessity of a teacher being present. Online courses may make a lecture available to virtually every interested citizen on the planet. But the academic development of the student and her achievement are fostered through other, additional processes that seem to work best in one-to-one and face-to-face settings. Educational technologies are currently not in a position to provide both aspects equally, as students and lecturers have realized during the Covid restrictions.

The challenge for these new systems is thus to combine the 'low production costs of mass production with the handcrafted effects of individual customization,' as the authors write (Susskind and Susskind 2016:130). Based on an extrapolation of existing trends, they are sanguine as regards future skills of AI machines in general: 'Our expectation is that systems based on Big Data techniques will often draw conclusions, offer advice, and provide guidance at the standard of human experts or higher' (Susskind and Susskind 2016:163).

But the examples do not always add up. They write that in ten years, 'robots have moved from making cars to driving them' (Susskind and Susskind 2016:167). Nowhere are driverless cars navigating the world 'in the wild.' Other examples are more convincing, as robots are around in warehouses, manufacturing, and other parts of the industrial world. They might also be able to decipher emotional states of human beings (and animals), which might make them useful in specific situations where they could provide 'companionship.' Skeptics will counter that pets and TV already provide such companionship; robots are not able to decipher emotional states, nor do they have social intelligence, more generally speaking (Frey and Osborne 2013).

The authors have a good point when arguing against the tendency of the professions to mystify their capabilities. This behaviour, which arguably dates back to the power of the priests as a social and professional group, involves the propagation of the belief that the professions are so complex and difficult to understand that no layperson could grasp their expert knowledge base and their function in society. This mystification is partly achieved by the use of jargon, and through the exclusion of clients and other interested parties. Susskind and Susskind use the sociological term 'social closure' to describe this mechanism, echoing the concerns introduced by Berger and Luckmann above.

But an important aspect is missing from this account. We should make a distinction between unwarranted professional closure and essential functions of the expert. While the first is undesirable the second is not. Essential functions of an expert are providing a guide for action that is not easily available to clients through better access to increased datasets. The ability of experts to provide this kind of practical advice, which is dear to the customer, supplies the base for their high esteem. As the authors rightly point out, there is an asymmetry involved between an expert and a non-expert in that the expert knows something the other does not. But as I have argued above, the knowledge is not only related to information about data, but to making judgements about conflicting data and values, a situation more and more commonly experienced.

Furthermore, developing AI systems which take over skills from human experts does not solve the problem of asymmetry. The mystification of the expertise is even enhanced in the case of AI, where the algorithms are essentially a black box, not open for interrogation.

In order to explore these issues in further detail, I dedicate the next three sections to a discussion of three examples of professional expertise: medicine, journalism, and law.

The case of medicine

Doctors as experts

In what sense are doctors experts? Based on the typology I am using throughout this book I think the following can be said. Medical experts are specialists who make and implement decisions as part of their professional jurisdiction. However, they also act as advisors to the client (the patient). Some medical experts are also seen as commentators in public debates.

Emphasizing these characteristics backgrounds the scientific aspects of the medical profession. It is true that some doctors are research active, and that in some cases new research influences treatment options. However, the proportion of research-active doctors among the population of doctors appears to be small. And the institutional context in which doctors operate usually demands recognition and approval of new treatment methods.

This means that medical practitioners as researchers are normally not in a position to apply research findings directly into medical treatment. They are only allowed to do this in the case of approved treatments. This is why I focus on the doctor as specialist, and not as researcher. As Ho et al. have pointed out,

> Healthcare is currently delivered by licensed professionals at accredited facilities or licensed premises, and designed to flow from a professional through approved channels to care recipients. The professional here is also the repository of knowledge, which underscores the notion of a fiduciary relationship where a clinician is required to act in the best interests of her or his patient.
>
> (Ho et al. 2019:330)

Citizens as medical experts

The term 'lay expert' is difficult to trace; apparently it was first used by Arksey (1994) with regard to the field of medical practice (although there are previous mentions in the 1980s, see Prior 2003). Arksey conducted a case study of patients suffering from a condition which was not recognized as an official medical condition, repetitive strain injury (RSI). As in many other similar cases patients realized how little their doctor knew about the condition. They organized themselves in support groups, challenging the medical establishment. They identified sympathetic and unsympathetic doctors—where patients were treated by an unsympathetic doctor they were encouraged to change to a doctor sympathetic to their needs. Unsympathetic doctors would, for example, claim that patients' complaints are psychological rather than physiological in nature (Barker 2008). Arksey maintains that the 'assumption that GPs are educated and patients are not is naive, and that under certain circumstances these attributes may be reversed' (Arksey 1994:462).

This claim became a hot topic in the STS literature, even if it was not attributed to Arksey. Before discussing the medical case in more detail I want to mention one of the classical studies about citizen expertise, Brian Wynne's work on Cumbrian sheep farmers. This was sometimes interpreted as showing that the farmers knew more or better than government scientists, when confronted with the consequences of the radioactive fallout from the Chernobyl accident in 1986 and subsequent government interference in their daily farming practices.

Wynne argued that while government scientists made authoritative statements about the radioactive decay in the farmers' soil, the farmers themselves knew something else about the nature of their herds and the requirements of animal husbandry. After all, the issue was how to solve the tension between radiation risks and maintaining the farmers' livelihoods. However, the government scientists and the farmers had specialized knowledge about different things, and both had their own interests. The local knowledge of the farmers was based on experience and self-interest, while the government experts relied on abstract models and an interest to reassure the farmers and the public, apart from keeping their monopoly status as provider of certified, reliable knowledge.

Perhaps the crucial point of Wynne's account is that the two forms of expertise, government soil, and sheep farming expertise, did not come together. They were not seeing eye to eye, and no attempt was made to bring both into a cooperative mode of addressing the problem they were facing. As we shall see, this stands in contrast to some examples in the field of medical expertise.

Returning to the example of medical expertise, Prior (2003) suggests that those who want to question the hierarchy, or asymmetry, between laypersons and experts, usually do not spell out what the nature of expertise is that laypeople allegedly possess. In some cases, laypeople are experts because they have personal experience. In other cases, lay experts are seeing eye to eye with those who have scientific training. And sometimes lay expertise can be seen as a property of patient groups that contain scientifically trained experts who serve as translators for lay concerns. Prior accepts 'that patients can have extensive knowledge of their own lives and the conditions in which they live.' She also recognizes 'that they can (and sometimes have to) turn themselves into experts in order to challenge medical hegemony.' But she disagrees that laypeople are experts: 'They are, for example, rarely skilled in matters of (medical) fact gathering, or in the business of diagnosis. What is more they can often be plain wrong about the causes, course and management of common forms of disease and illness' (Prior 2003:45).

True, not all lay people are experts, the question is if they cannot be experts as a matter of principle. It is also true that they can be plain wrong, but this applies to professional and scientific experts as well.

Rabeharisoa and Callon (2004) take up a similar example as Arksey, a disease that was neglected by the medical establishment. They present the case study of the French Muscular Dystrophy Association (Association Française contre les Myopathies, AFM). The authors write: 'Numerous accounts describe the vicious circle of abdication and ignorance: when a disease is unknown the professionals turn away from it because it highlights their powerlessness. This disinterest, in turn, maintains the state of ignorance because it paralyses all efforts at carrying out research on these diseases. In such situations, the only knowledge on the pathology is that developed by the patients and shared with their families' (Rabeharisoa and Callon 2004:145).

The AFM helped to collect information on the disease, which they call, echoing Marx's famous words, 'a primitive accumulation of knowledge.' The patients' experience served as the starting point for the development of a body of knowledge which then could serve as the basis for the systematic production of expertise. The AFM was already established in 1958 and was instrumental in supporting patient organizations, providing financial support up to research, exerting pressure on public decision-makers, and organizing patients' participation in clinical tests. It provided a platform for interaction between patients, scientists, and clinicians. This was done through organizing discussions that enabled mutual learning. They created 'hybrid fora' where groups of diverse origins could mix. 'They interact, discuss and negotiate, seeking answers to questions on the research to undertake, the evaluation of results obtained, and the programmes to support. The main characteristic of these forums is that they organize collective experiments: experts are in constant dialogue with lay people, and actions are decided together after joint analysis of the results obtained.' These fora established 'a rich network of diversified relations between specialists and patients, caught together in the production of the history of the disease' (Rabeharisoa and Callon 2004:148–149).

The term 'collective experiment' is remarkable, and the fora described here seem far from normal procedures. This sounds more like an ideal setting for co-production, perhaps comparable to the case of AIDS activist groups described by Epstein (1996). The authors take this a step further, and detail how patient groups in the case of AFM develop, collect, and interpret their own research data:

> 'Patients and their families play a crucial role in the primitive accumulation of knowledge and know-how on their diseases. .... Not only do they act as a driving force in the recording of information on patient populations; they also, by means of proto-instruments, put themselves in a position to produce formalized knowledge and practical know-how, collectively and in close collaboration with specialists. The setting up of tripartite structures (patients, clinicians, biologists) for work and discussion, enables them to build on learning and to circulate and evaluate results. Patients thus become both subjects and objects of research.' (Rabeharisoa and Callon 2004:157).

In this account, lay and expert knowledge constitute each other, through a process of cooperation and mutual learning.

These two examples from medical practice are similar in that patients rally for better medial support from the health system regarding a medical condition about which not much is known. There is a more general lesson in this. Patients have challenged the medical establishment, and a similar logic can be observed with regard to 'established' diseases that have been known for a long time. In some cases patient groups have influenced the process of knowledge creation itself.

Coming back to the difference between Wynne's sheep farming case and the examples of the patient groups the question arises if the difference lies in the difference of the two policy domains or if other explanations are needed. A cursory review of the literature suggests that social change is the main factor here. Since the Chernobyl accident in 1986 several high-profile public debates have taken place, in various countries (see Grundmann 2008 and the cited literature). GM food, BSE, mobile phone radiation, and nuclear power plants are only some prominent examples. Several, diverse academic paradigms have emerged since then, among them The Public Understanding of Science, Post-normal Science, Co-production of Science and Society, to name a few.

However, in contrast to patient groups there seems to be no evidence for cooperative efforts in the field of environmental policies. The notion of collective experiments in the sense used by Rabeharisoa and Callon does not fit here. As Lezaun and Soneryd (2007) have pointed out with regard to adverse effects of GM technology and mobile phone radiation, the nature of the debates was a public consultation during which governments elicited concerns and issues 'from below,' rather than entering a dialogical mode in which collective experiments and cooperation would be aimed at. This suggests a power asymmetry where one side is in charge and at liberty to use the results of the consultation as it sees fit.

Environment means territory, and the state is in charge. Health means personal embodiment, and the patient owns it. Health professionals may try to take charge, but this often causes a tide of activism going against such attempts. Covid-19 and the public health imperatives are the latest prominent example. Government, by contrast, operates at a different level. Being the site of ultimate decision power in society it is always aiming to preserve its power base. Collective experiments in this case usually take place on another level, the level of politics and elections.

These brief remarks may point to a different logic in the fields of health and the environment, such that patients are able to change medical practices through interactions with the medical profession. Environmental policy on the other hand is from the outset a politicized field where governments want to keep control, and where environmental activism articulates itself through protest. This could only change where environmentalist activists develop new practices in collaboration with other social groups (Papadopoulos 2018). In analogy to the health examples discussed above, this would mean that activist groups establish new modes of production on the ground, circumventing the political power game. The ideas of utopian socialism might make a new appearance.[1]

The other main mechanism for changing environmental practices is the application of technological solutions, provided by the engineering professions, and a network of private and public organizations. Here societal expectations and responses are diverging. While some place their hopes in these solutions, others decry them as misguided techno-fixes.

Above I have identified a pincer movement which advances against established forms of expertise, citizen expertise and AI. The following section will explore the role of AI as a form of expertise in the filed of medicine.

The challenge of AI

A recent summary of the role of AI in medical settings identifies several issues that are important when considering the potential and risks it may engender (Ho et al. 2019). Currently, 'clinical medicine increasingly reflects a shift-based model, where fewer clinicians chart diseases from their presentation through their ultimate outcome.' AI could change this setting and AI applications could disrupt such healthcare practices. Nevertheless, a clinician would still remain 'at the centre in attribution of ethical and legal responsibilities, in payment arrangements, and in organizational setup.'

A special problem is the ethical aspect of data protection: 'AI may curtail a patient's exercise of her or his right to privacy and confidentiality as ML (machine learning, R.G.) analysis will require the patient's personal information to be recorded' (Ho et al. 2019:330). Above I argued that health is embodied and the patient is in charge. AI will create a fault line vis-à-vis patients' expectations and their rights.

Before proceeding, a conceptual distinction is in order. Sometimes the term medical robots is used in the literature. This must not be confused with automated systems which are based on machine learning and big data. As an overview paper states, '[r]obotic systems for surgery are computer-integrated surgery (CIS) systems first, and "medical robots" second. In other words, the robot itself is just one element of a larger *system designed to assist a surgeon* in carrying out a surgical procedure that may include preoperative planning, intraoperative registration to presurgical plans, use of a combination of robotic assist and manually controlled tools for carrying out the plan, and postoperative verification and follow-up' (Taylor and Stoianovici 2003:765, my emphasis).

Interestingly, automated systems which had been introduced in a few instances are now seen as undesirable, as one commentator pointed out: 'In surgical robotics, there has been a trend away from autonomous or even semiautonomous motions, and toward synergistic manipulation and virtual fixtures. Thus, the robot acts as a guidance tool, providing information (and possibly a physical nudge) to keep the surgeon on target' (Beasley 2012:10).

Let us consider some of the examples in greater detail. IBM has developed an online medical expert system known as Watson (Ferrucci et al. 2010). It is being used for the diagnosis of cancer and for the development of treatment plans. The advantage of this system is to improve access to the largest possible database and to improve accuracy in knowledge. The first aspect is important because of the sheer volume of research publications which makes it impossible for the medical professionals to keep up to date: A new paper is published every 41 seconds, as Susskind and Susskind (2016:48) claim. If only 2% of the newly published medical literature was relevant for a doctor, it would take a human 21 hours each day to read it. Watson might help improve the knowledge base for making better and more accurate diagnoses and improve treatment.

While this sounds compelling, there are various problems. As Gerd Gigerenzer has pointed out, in order to make successful predictions we need a good theory, reliable data, and a stable world. Machine learning algorithms typically rely on correlations, and do not involve causal theories. As Chris Anderson, chief editor of Wired magazine, famously said, 'Forget taxonomy, ontology, and psychology … Petabytes allow us to say: "Correlation is enough." We can stop looking for models. We can analyze the data without hypotheses about what it might show. We can throw the numbers into the biggest computing clusters the world has ever seen and let statistical algorithms find patterns where science cannot … With enough data, the numbers speak for themselves' (Anderson 2008).

This view may be extreme but is shared more widely. It forms part of a narrative which professes techno-optimism but ultimately deals with issues of control and responsibility. What is interesting in this context is the epistemology underlying the big data enthusiasts which goes against everything a scientist will have been told in their education. That correlation is not to be confused with causation is the first article of any serious scientific endeavour. This casts an interesting light on the question of boundary mechanisms: who is defining what good science looks like in the field of big data and AI? Reliable data is often impossible to obtain, and data-mining does not necessarily improve the situation. The so-called replication crisis in several disciplines, medicine included, is an indication of the problem.

A serious problem is that AI may be very good in experimental settings, but not in the real world (Gigerenzer 2022). Machine learning systems, even those based on deep neural networks, are trained on data which is cross-validated. The learning refers to pattern recognition on datasets that are supposed to represent the real world. But AI systems cannot access the real world, or the relation between reality and its representations. All it can do is relate data to data. This sometimes leads to stunning results, especially considering the speed at which enormous amounts of data are analyzed. But the results need to be tested in experimental ways, 'in the wild.' This is where problems and costs tend to multiply.

There are examples where technical systems have proven to be useful. In the example of the pacemaker, medication has been replaced by a medical implant. In the case of the insulin pump the cumbersome process of individual blood testing and insulin injection has been taken over by a medical device that is more reliable than human intervention by simply injecting the correct dose of insulin, depending on the blood sugar level of the patient. Both machines are little cybernetic devices that respond to the state of bodily functions, constantly adapting their output. However, the use of the term robot, or AI technology in these cases would be misplaced. The devices mentioned execute an action based on a simple feedback loop. They operate as a trivial machine (Foerster 1981) where every input corresponds to a predetermined output. AI applications are not trivial machines and thus unpredictable. We only consider their service where we trust that the output is safe, valuable, trustworthy, or desired in some other way. Most medical applications are not autonomous agents but subordinate to the surgeons' decisions and actions.

The companies developing such devices are pushing their products in subtle and not-so-subtle ways. When assessing them one needs to be careful not to fall for the marketing hype. The example of IBM's Watson shows that there is more hype coming from marketing than a working product. Not only did Watson fail to distill important information from the medical literature but it also failed to mine information from patients' electronic health records (Strickland 2019). One application, called Watson for Oncology allegedly recommended 'unsafe and incorrect' cancer treatments (Ross and Swetlitz 2018).

In 2020, the Financial Times reported on the prospects of big tech companies to enter the healthcare sector. Using machine learning technology seemed to provide the tool to reconfigure and reconceptualize healthcare services. However, it appears that such initiatives run counter to the needs and expectations of the medical profession and patients. What is more, the tech companies have over-promised and under-delivered, as was the case with IBM's Watson, which is supposed to cure cancer. AI may be available for diagnosis but it is not for treatment. 'There are too many treatment options available, and subtle clues about patients that only experienced practitioners can identify' (Waters 2020). Machine learning technology as such does not lead to a radical change in the health sector. The companies developing it seem to realize they need to build relationships with relevant stakeholders first and foremost. The issue of patient data, its sheer volume and sensitive nature, is a particular problem.

A chief scientist at Microsoft is quoted as saying, 'Tech companies ... have often underestimated the day-to-day needs of healthcare professionals, failed to create services that are tailored to specific medical settings, and ignored the many "human factors" that have limited adoption of new technologies' (Waters 2020).

Even more explicit and damning is Oren Etzioni's verdict: 'IBM Watson is the Donald Trump of the AI industry—outlandish claims that aren't backed by credible data ... Everyone—journalists included—know[s] that the emperor has no clothes, but most are reluctant to say so' (cited in Brown 2017).

Narratives and counter-narratives abound in the field of AI. Nevertheless, they can be summarized into broad genres, utopian and dystopian, reassuring and alarming. The fact that AI does not deliver what its creators promise is seen as a temporal lag by its developers. Given time, the technology will catch up and live up to the promise. The structure of the narratives is slightly different compared to climate change and Covid-19. The elements of plot, heroes and villains, and moral are more in the background, and the questions of control and responsibility become much more prominent. Still, the reassuring and alarming tone can be sensed widely.

Journalism as professional expertise

Journalism dates back to the 17th century and is a professionalized activity. It has, as other professions, come under pressure with the ascent of the knowledge society and the internet. Citizen journalism and AI challenge the traditional model of journalism.

Journalists do not reflect social reality like a mirror; they select information and present it in ways to attract an audience. While they are never 'objective' they can try to be impartial. They can also be openly (or tacitly) partisan. We see huge variations over time and across political cultures, as continental Europe and the United States indicate. Writing about Europe, Fürst et al. (2015:334) claim that

> The beginnings of journalism were shaped by the journalistic role conception of providing impartial mediation and a platform for social discourse. Historical studies on journalistic show that German-speaking as well as English-speaking journalists of numerous newspapers since the seventeenth century were committed to the concept of impartiality. It included the following dimensions: pluralism or diversity of voices, standpoints, issues etc.; separation of sources' messages and journalists' opinions; transparency of sources, so that their interests, relevance and representative function could be assessed by the audience; and preservation of original sources' meaning to avoid contortion of their statements and arguments. Impartiality in this sense was not only professed by journalists, but it widely guided their news coverage as studies of the first two newspapers in the seventeenth century as well as of many other newspapers in the eighteenth and nineteenth centuries indicate.[2]

Now consider the account given by Schudson (1978) about the USA:

> Before the 1830s, objectivity was not an issue. American newspapers were expected to present a partisan viewpoint, not a neutral one. Indeed, they were not expected to report the "news" of the day at all in the way we conceive of it— the idea of "news" itself was invented in the Jacksonian era.

> 'At the turn of the century there was as much emphasis in leading papers on telling a good story as on getting the facts. Sensationalism in its various forms was the chief development in newspaper content.' This only changed with the advent of news wires, such as Associated Press in 1848, as Schudson explains: 'Since the Associated Press gathered news for publication in a variety of papers with widely different political allegiances, it could only succeed by making its reporting "objective" enough to be acceptable to all of its members and clients. By the late nineteenth century, the AP dispatches were markedly more free from editorial comment than most reporting for single newspapers.'
>
> (Schudson 1978:4)

This interesting difference between Europe and the US is due to the party political funding of early papers in the US, that is, in the period before 1848. During that time newspapers were read by a small elite. This only changed with the expansion of the middle classes and the newspaper market which led to an inclusion of more citizens into an emerging public sphere. As Kaplan (2001:10) points out, 'this mass inclusion and accompanying institutional and cultural change were the social background for a new, specifically modern type of newspaper. The change created the (economic) resources and (cultural) demand for the modern press as the purveyor of a distinctive, novel commodity: the news.'

The economics of newspapers applies across the Atlantic, and the printing press made printed news affordable. Likewise, state censorship and control of newspapers was an issue, and it still persists today. While several countries have adopted and implemented legislation granting freedom of speech, not all have. The governance of the media is a constant concern, world-wide. It is unrealistic to assume it could be independent of economic or political constraints.

To provide 'the truth' is perhaps demanding too much from journalists. But if they can provide an impartial account which stands up to scrutiny much has been achieved. This is arguably under threat with new forms of citizen journalism, especially on social media platforms. British journalist Jonathan Freedland shows what is at stake. He accuses two senior UK politicians of lying, and some social media publications of spreading the lies. Both could only have happened because no journalist dared to oppose the false claims, or was not given the platform to contradict prominently:

> 'Fact checking is laborious, tedious and time-consuming, especially compared with the brio that can be generated by a sweeping (but false) assertion. ... Perhaps that was always so. More significant is the shift in media culture. Technology now allows politicians to communicate directly with their followers, with no need to transmit their claims through the fact-checking filter of a news organization. ... voters can easily get the entirety of their information from sources tailored to reflect their own

> views back to them. That, in turn, makes them ever more unwilling to accept contradictory facts.' The editor of the Washington Post understandably laments 'What has taken hold is an alternate reality, a virtual reality, where lies are accepted as truth and where conspiracy theories take root in the fertile soil of falsehoods.'
>
> (Freedland 2016)

Bypassing the quality control checks of professional journalism has become easier with the affordances of social media. Politicians can make claims without being interrogated. The problem clearly existed in the past, when leaders used radio, television, or press releases. The crucial problem is one of political power, and its accountability. If politicians are not held to account for their statements and actions this creates a problem for legitimate rule in a democracy.

This is where journalism has a crucial role to play. However, it faces several challenges. First there is the pressure to cut costs, especially after the financial crisis of 2007/8 and the shift of advertising from newspapers (both print and online) to online platforms (Sridhar and Sriram 2015). Then there is the rise of 'citizen journalists,' writers who have no professional training in the trade. They range from random eyewitnesses to university professors. All are invited to write, or comment on traditional or non-traditonal online media platforms, or to produce their own content in self-managed formats, such as blogs. Finally, there is the increasing application of automated software-based systems which allow to collect, retrieve, organize, and summarize news. Contrary to medical robots, journalist robots are a real existing technology, in the sense of autonomous agents.

What kind of expertise do we see in journalism? Compared to medical experts journalists have little jurisdictional power and their relationship with a client is very indirect. Journalists exemplify the role of the commentator. They provide news to an audience, but can also act as issue advocates, providing guidance for action. With the rise and automation of news sources the demand for writing news will decline and the demand for advice could increase. Journalists may find themselves shifting towards the role of partisan commentator, providing guidance for their audience. This development could blur the lines between journalist and advocate, with many members from civil society, business, and political parties offering their comment and advice to a broader audience. This would signal a return to the situation described by Schudson above, where partisan newspapers dominated the scene in the US around 1830.

There are some institutional barriers to this development. Perhaps the most important one is the constitutional duty of public news broadcasters to inform in a non-partisan way. In the US, a minimalist regulatory framework incentivizes private broadcasters to compete for viewers, and US television news programming have not been subject to the same regulatory requirements compared to many European broadcasters (Cushion et al. 2012). In the UK, for example, since 2011, commercial broadcasters have been regulated by the

Office of Communication (Ofcom) while the BBC is regulated by the BBC Trust. The latter defines itself as an interventionist watchdog, claiming from a public service perspective to 'have considerable power to wield on your behalf – and when we need to, we act quickly and decisively in your interests' (BBC Trust website).

The BBC 'is required to ensure that nothing is included in its programs that offends or is against good taste or decency, or is likely to encourage or incite crime or lead to disorder, and that any news given (in whatever form) in its programs is presented with due accuracy and impartiality' (Brants and Bardoel 2008:475–476).

By contrast, Ofcom was conceived as a 'light touch' regulator. All UK broadcasters have to be impartial in their reporting and have to carry news and current affairs programmes, also on local and regional levels, which require investments that some providers resent. The push to further deregulation is evident (Cushion et al. 2012).

Overall, the changing media landscape holds various possibilities for the future. The governance of media outlets, the question of censorship, and of impartiality will be high on the agenda. Narratives about these developments are evident in most countries, reassuring and warning us.

The rise of citizen expertise

Hayley Watson (2011) uses the term 'citizen journalism' to denote the involvement of citizens in news production: 'Citizen journalism involves the activities of members of the public in contributing to the production and distribution of news items in society.' But different authors understand different things by this.

In one version, 'participatory journalism' includes collecting, reporting, analyzing, and disseminating. This is said to involve 'very little editorial attention' (Bowman and Willis 2003). In this view participatory journalism is a bottom-up phenomenon, which originates from 'simultaneous, distributed conversations' that can either gain popularity or gain very little attention from Internet audiences (Bowman and Willis 2003: 9, as quoted in Watson (2011)).

Some commentators think that citizen journalism is different to professional journalism. The latter discloses news, whereas the former is mainly re-telling stories. However, as Goode (2009:1290) argues, also professional journalism 'is in no small measure a craft of re-telling stories rather than simply disclosing them. One may consider the common practice within mainstream journalism of reworking press releases and agency feeds.'

He suggests a broader framework of mediation in which both professional and citizen journalists can be located: 'Rather than treating news as revelation, then, it is possible to situate citizen journalism within a framework of mediation that can account for a wide spectrum of news-making practices, from activists blogging about local public body corruption, through cell phone photojournalists, to taggers who contribute to shifting memes of public discourse through the simple act of labeling news stories already in circulation' (Goode 2009:1291).

There is a democratic deficit in professional journalism and mainstream media as there is no transparency of the story selection process. Citizen journalism can counterbalance or challenge those selections. However, this will not do away with the agenda setting power of the big news providers: 'Such institutions still break and frame a large proportion of the news stories circulating through the online sphere and this is unlikely to change in the foreseeable future. But those institutions must now vie for attention in competition with a diverse range of alternative news sources, from hyperlocal sites to unofficial and untamed celebrity gossip sites' (Goode 2009:1292–1293).

This means that the ecosystem of news gets enlarged, and there is more competition for scarce attention. A lot of citizen journalists' output will be simply ignored or forgotten. But some stories will be elevated to the highest levels of attention: 'Stories, once online, confront various possible fates: they may be more easily buried in this vast new attention economy if they do not capture the imagination quickly and strongly enough; or they may be amplified, sustained and potentially morphed as they are re-circulated, reworked, and reframed by online networks' (Goode 2009:1293).

Goode wrote these comments in 2009, before the widespread adoption of social media online services such as Facebook, Twitter, etc. His empirical examples are about sites that have sunk into oblivion. Nevertheless, the observations are still a useful first step in approaching the phenomenon of citizen journalism.

The same can be said about a study of the 2010 cholera outbreak in Haiti, in which Krajewski and Ekdale point out that '[c]itizen journalism … often is praised for its ability to act as a democratizing agent, elevating the voices of those excluded by traditional media gatekeepers … Although studies have found that citizen journalism can provide an important counternarrative to traditional media, citizen journalism does not exist outside of hegemonic power structures' (Krajewski and Ekdale 2017:230).

Citizen expertise in the realm of journalism can manifest itself in different ways. Activity on social media platforms such as Twitter or Facebook is one example. People share pictures, videos, statements, jokes, and petitions, some of which they have created themselves. Users also set up their own managed content, as with blogs or personal websites. Finally, they contribute to mainstream media providers who have an online platform. This can take the form of providing stories through contributing text or pictures, or providing comments to articles provided by professional journalists.

The now-defunct CNN iReport was one such instance. On its website it boasted, 'We're excited to announce that the CNN iReport community has reached one million registered members. This is a major milestone, so we want to start by thanking you for choosing to spend your time with us and for sharing your stories with the world. It's really cool to think that our community has the population of a good-sized city and that iReporters have posted stories from every country in the world.'

Problems with vetting news reports from 'the audience' were plenty. The most famous is when a report that Steve Jobs had suffered a severe heart attack in 2008 led to a fall in Apple's share price. CNN withdrew the service in 2015 and has changed its approach. Problems with news vetting were not even the most prominent reason for the change but the lack of readership. CNN moved to establishing a presence on social media where their readers are, mainly on Facebook and Twitter (https://digiday.com/?p=146002).

The BBC has been using user-generated content (UGC) for some years now. Lisette Johnston, an academic at City University who also works as senior producer for BBC World News TV, has published research on her own organization, using the example of UGC in the Syrian conflict where citizens have provided video material. Her conclusion is that journalists have to be present on social media platforms as newsgathering sources and places to seek out eyewitnesses or footage. 'This means journalists have had to become more flexible and adaptable in digital spaces, perhaps having a greater presence on certain social media platforms as well as becoming more proficient in understanding technology which will help them track down uploaders, or content. With regard to UGC specifically, journalists have been challenged as they have had to adapt quickly and learn how to treat this content' (Johnston 2016:907).

Commentators are in two minds about such developments. The opening of mass media platforms to non-professional journalists is like the emergence of field expertise in other areas, such as health and medicine. The emergence of field expertise means critical questioning of established, professionalized expertise. On the other hand, the same process means that some essential standards of quality might be lost. But what exactly essential standards are is a matter for debate, as we shall see.

David Brake (2014:604) celebrates the benefits of online content creation (OCC), 'for those who create should also be considered now that the tools to distribute creative content are (in principle at least) available to a much wider potential population than ever before.' He argues that the 'production of culture in everyday life is beneficial to the individual creators, making them more satisfied by giving them a project they can share with others, and also benefits society at large by helping to build social capital and giving those who may not be accustomed to wielding power and influence a sense of their ability to make a difference. For all these reasons, broadening the base of people who know about, can access, and can effectively use online content creation tools is an important and underconsidered goal.'

On the other hand we have witnessed the emergence of automated software-based systems which allow to collect, retrieve, organize, and summarize news. However, as Splichal and Dahlgren (2016:10) point out, the prospect of automation does not cause the trend towards de-professionalization; it merely exacerbates it: 'The true cause of de-professionalization and structural unemployment in journalism is not the automation of journalistic activities but rather the fundamental transformation of the essence of journalism and its degradation to information work, which is part of a broader process of growing structural unemployment caused by automation of work in both stagnant and progressive sectors.'

By stagnant and progressive sectors they mean two types of economic sectors, one in which labour productivity remains stable, and one where it increases. According to Baumol and Bowen (1965), the arts are the prime example of a stagnant sector. Little technological change happens and therefore productivity does not rise. Wages thus account for a substantive ratio of the investment. Journalism has long fallen into this category, but what happens in the age of the internet, speech and text recognition, and social media?

The point has been illustrated as follows:

> The internet brought mass media as a specific sector of cultural industries closer to the 'progressive industries' with a higher productivity growth, particularly by the successful commercialisation of machine learning, implemented by the major global digital media players … The open-access platform EventRegistry (2015) … finds and displays articles by crawling online news sites, analyses news items available online, identifies groups of news items reporting the same events in different languages and extracts event's core information, such as geographic location, date, actors and actions. A user interface allows users to search for events using extensive search options, to list, visualise and aggregate search results in tag clouds. Wordsmith platform patented by Automated Insights (2015) is able to analyse 'big data', identify relevant insights and turn them into reports in natural language. Such automation technologies are used by Bloomberg News, Google News, Associated Press and many other news corporations.
>
> (Splichal and Dahlgren 2016:9)

How does the sector of journalism respond in economic terms? The authors think that it neither corresponds to the logic of the progressive sector, where new technology leads to wage increases, nor to the logic of the stagnant sector, where new technology leads to the recruitment of more staff. Their conclusion is pessimistic: 'Changes brought about by automation are part of the creative destruction of journalism making it much more productive but less and less – journalism.'

The robot journalist

'Imagine an organization has just hired a new writer. Her boss discusses the topic of coverage with her and passes along a style guide. She writes a bunch of samples. She gets feedback. She writes some more. At some point the feedback stops because she knows what she's doing. According to Kris Hammond,

Narrative Science's technology head, that's exactly how his company's technology, called Quill, works. 'It just happens to be the case that it's a machine,' he says (Quartz, October 9, 2013, cited in Carlson 2015:425).

This quote illustrates the principle of machine learning. Software systems have been devised that, after some training, can analyze news data and write stories. Two high profile companies are the above-mentioned Narrative Science, and Automated Insights. The latter advertises its services as follows: 'Spin thousands of articles with no extra man power fast, cost-efficiently & multilingual. Unique content – High quality – 110 languages.'

The software emulates the style of human writing, thus trying to hide the features of non-human writing. The results in some areas are good enough to be published. Here is a list of comments about examples from recent years:

> The stories drew qualified praise, including: "it might not be riveting, but it is perfectly readable" (*The Sunday Times*, May 20, 2012). Peter Kafka (All Things D, February 16, 2012) added: "The result isn't elegant, but it gets the job done, in a brute force sort of way." Other assessments made clear that the stories lacked the quality of the best writing, but were still serviceable as news accounts: "Though these pieces lack the verve of, say, Chuck Klosterman's sportswriting, the highly customizable platform does adopt a sports fan's idiomatic shorthand" (The Atlantic, April 12, 2012). Fast Company (June 19, 2012) added, "The phrase 'convincingly human' has probably never been used by the Pulitzer Prize committee, but it's good enough when it comes to analyzing large data sets, or the earnings reports that Narrative Science files for Forbes.com."
>
> (Carlson 2015:424–425)

Carlson comments: 'The phrase "good enough" in the last quote usefully captures the recurring sentiment that automated journalism is capable of meeting minimum expectations that news writing provide clear and accurate information. But it also invokes what Robert Capps (2009) labelled the "good enough revolution"— the emergence of technology whose success is not due to high quality but in being functionally adequate and inexpensive.' The question about essential standards arises and here is a suggestion that 'good enough' is the new measure for journalistic quality. It is difficult to reconcile this approach with notions of 'quality' journalism or 'journalistic excellence.'

Accuracy and objectivity are invoked as main criteria when evaluating information, knowledge, or news. The appeal and performance of AI is there for all to see, and the prospect seems bright. The chairman of Narrative Science went as far as saying that 'within five years a computer program will win a Pulitzer Prize—and I'll be damned if it's not our technology' (New York Times 10 September 2011, cited in (Carlson 2015:427). We may dismiss this as hype and his prediction has not come true yet. But who would want to bet against it?

One study has tested samples of text written by journalists and robots on human readers. They could not tell the difference between the two: 'As far as this study is concerned, the readers are not able to discern automated content from content written by a human. Some aspects of quality, such as being clear and being pleasant to read, received a slightly higher score for human-written content, but others, such as trustworthiness, informative, and objective, were higher for the automated content' (Clerwall 2014).

In a different study using a much bigger sample these findings were supported: 'First, computer-written news tends to be rated higher than human-written news in terms of credibility. Second, news consumers get more pleasure out of reading human-written as opposed to computer-written content. Third, differences in terms of perceived credibility and expertise tend to be small' (Graefe et al. 2018:604).

Traditional journalism faces a problem of trust. All media reports are biased to some extent, errors in reporting are normal, and fraud is not unusual. Invented stories existed long before fake news became a topic. Against this background robot journalism is put in front of us as a proposition, a proposition of more accurate and more trustworthy reporting. As someone said in the business magazine *Fast Company*, 'Good journalism isn't about writing like a human. It's about trust. And as trust in conventionally authoritative sources continues to erode, Narrative Science's robots may be lying in wait to pick up the slack' (cited in Carlson 2015:427). The selling point of robot journalism resides in its 'algorithmic objectivity' compared to the shortcomings of human journalists.

It may sound eerie but the old hermeneutic adage 'the text knows more than the author' seems appropriate here. From the earliest written traces of human civilization there is a deep insight that texts, and data, are there to be interpreted. Any specific meaning given to them is the privilege of an author, who puts their own spin on it, and of the reader who wants to perceive specific messages and to interpret texts in unique ways. In this sense the AI revolution could be seen as providing an additional layer of interpretation of data, not mediated by human intention or bias. True, the machine learning algorithms are written by humans, and therefore make the outcome potentially biased by replicating the programmers' bias.

An advantage of automated journalism could be that it takes up routine tasks, which will free up time and resources for human journalists to do investigative and in-depth analysis. Robots write news about finance, sports, and weather, while humans write background stories, deep analyses, and witty comments. However, the increasing trend to produce personalized news implies a deep fragmentation of the public sphere through the creation of filter bubbles (Graefe et al. 2018). AI applications will be even better at this, leading us ever deeper into echo-chambers.

If algorithmic journalistic authority resides in accuracy and objectivity, defenders of human-produced news emphasize that journalistic authority derives from something more than delivering objective information about the world. Humans write with wit and empathy, and the writing involves an emotional quality, something which AI systems do not master (Frey and Osborne 2013). It may be that a hidden contest between both logics of authority is already under way. This should give rise to public debates about what we expect from journalism and publicly trusted knowledge, as one author points out: 'To fulfill their aim of providing knowledge to the public, both systems must seek recognition as legitimate knowledge producers. As the technology improves and the umbrella of automated journalism grows, these two logics will be put into conversation with one another in ways bound to raise larger questions about what constitutes quality public knowledge and how it should be created' (Carlson 2015:428–429).

News providers could be made to reveal which of their content is authored by human journalists and which by AI systems. In 2019, *The New York Times* said it had no plans for machine-generated news articles, but this may change. Bloomberg is said to use robots for about a third of its content, and the Associated Press, *Washington Post*, and *Los Angeles Times* have used them for the coverage of sports events and earthquakes (Peiser 2019).

Fake news

Although historically not a new phenomenon, this has received much more attention with the advent of electronic networks and the rise of social digital media. Fake news is as old as news communication, even before the advent of mass media. In recent years researchers have studied various aspects, such as the availability of a plethora of information 'on the internet' which cannot easily be assessed for veracity. In a competitive media landscape the outrageous and sensational news will always attract a larger audience, and this has only been exacerbated by 'clickbait.' News headlines have been part of the same strategy for a long time.

What seems to be attracting most attention in current research is the ability to detect fake news through automated analysis of news. At the time of writing the most cited paper (Lazer et al. 2018) calls for an interdisciplinary approach to combat fake news, and this involves detection of fake news via algorithms which would be used to develop a quality star rating system on search engines and social media platforms.

Reuters has developed a piece of software which is able to detect news events on Twitter and to assess their veracity. It says that this system has 'been implemented at large scale and successfully discovered breaking news faster than traditional journalism' (Liu et al. 2016).

To sum up this section, it can be said that journalism is an example of a professional domain which has been strongly affected by field expertise and autonomous technology. In contrast to medicine, where robots assist the surgeon, in journalism robots write stories. Journalism is a field of expertise which provides news and comments to an audience. It has a low degree of abstract knowledge which means that it has less professional power than medicine. News gathering and dissemination is a job perfectly suited to electronic systems. In theory comments could be written by AI robots as well but perhaps the audience prefers to get comments from specific commentators—although even here a competitive process could set in.

Similar to the case of medicine, influential narratives by companies developing AI tools are giving a positive outlook, even promising a better journalistic product as a result, less biased and more accurate. On the other hand, sceptics point to the dangers of technological black boxes which are out of the journalists' and readers' control, albeit deployed by private companies. Issues of trust, control and responsibility come to the fore.

The case of legal practice

Like the medical expert, the legal expert should be seen as a hybrid between specialist and advisor, with some scientific element. As judges they have jurisdictional power to make and implement decisions; as lawyers they find themselves in a relation to a client who wants advice on a course of action to be taken. And as scientists they assess previous cases and legal frameworks.

Here I will argue that at first sight law presents a 'hard case' of professional expertise that has largely managed to resist change that has befallen other fields, like medicine or journalism. Neither AI systems nor field expertise seem to have made many inroads into law. However, upon closer examination one finds that law is rather similar to medicine in that lay participation can be identified, and technical expert systems have been developed to assist with juridical decision-making.

Citizen expertise in law

In contrast to medicine and journalism where we can identify citizen participation or the emergence of lay expertise, law seems to pose a different case. Despite the fact that legal decisions are sometimes jarring with public sentiment, the barriers to entry into the core of the business seem to be much higher. The law-making process on the other hand is done by parliament (at least in democratic societies), where elected politicians are in the driving seat. In this sense, laypeople do have an input into the legal process. They can participate in elections, or can appeal against court decisions through petitions.

However, there are two more ways in which law can be affected by citizens: one is the citizen participation in juries, and, in some cases, the referendum. In the case of juries, the involvement of citizens with the legal system is on the level of legal applications, not the production of law (although there may be room for this in case law). Referenda are tools which can be instrumental in the creation of new law. Both of these tools can be seen as enhancing democracy or providing opportunities for populist mischief (Brooks 2004). Jury trials are used in England and Wales, and other common law judicial systems. They are also used in many civil law countries for criminal cases (Hans and Vidmar 1986). Some legal theorists and practitioners are skeptical about their function in a democracy. For example, Judge Jerome Frank said: 'While the jury can contribute nothing of value so far as the law is concerned, it has infinite capacity for mischief, for twelve men can easily misunderstand more law in a minute than the judge can explain in an hour.' Or take Penny Darbyshire: 'What business have the jury to be rewriting the law? ... The jury is an anti-democratic, irrational and haphazard legislator, whose erratic and secret decisions run counter to the rule of law' (both cited in Brooks 2004).

As we can see, the public participation in the making, and application of law raises fundamental questions about the proper role of the demos in a democracy and how it is aligned or distinguished from the role of experts.

> The common law practice of trial by jury, whereby citizens decide cases arising from the communities in which they reside, began millennia ago, but the requirement that juries be impartial did not come about until the late Middle Ages. As anathema to justice as it may sound, early courts considered a juror's familiarity with the facts of a case to be an asset at trial and even preferred to impanel those with first-hand knowledge of the events at issue.
>
> (Hoffmeister and Watts 2018:261)

The ubiquity of news in the age of smartphones makes it difficult to control the jurors' requested allegiance to the court's order not to seek extra information relevant to the case, additional to the information provided in the legal setting.

Such personal curiosity and active involvement have been evident in the past and the availability of online news archives to jury members could be interpreted as a 'return to the historic role of jurors as active, self-informing participants in the process, rather than passive fact finders. Perhaps the concept of trial by jury finally has come full circle' (Hoffmeister and Watts 2018:261). The difference is that today the juror can get information that is not limited to geographical proximity.

Jurors are able to get access to information about the defendant (either inside or outside of the courtroom), thus undermining a fundamental principle of the jury system. Currently there exists no technological solution which could prevent this from happening. This has given rise to serious concerns about this particular form of citizen participation. As Hoffmeister and Watts (2018:268) put it,

> Internet and social media prominence has made it more likely than ever that (a) unbiased jurors will be hard to come by; (b) impanelled jurors will

be tempted to seek and consider outside information; and (c) unsanctioned juror communications with other jurors and nonjurors, both inside and outside the courtroom, will threaten the sanctity of the institution.

Note the use of the term 'sanctity of the institution,' which indicates that the legal system and its key institutions, including trial by jury, are seen as standing above the profane business of common social affairs. Googling a person has become a widespread habit but such behaviour is not acceptable for jurors. The holy institution of the law clashes with the everyday of socio-technological networks and citizens' expectations. This stands in marked contrast to other social domains, including medicine and journalism.

The law in modern states is supposed to work according to its own principles and be shielded against external influence. Such influence may come from interested parties who are privileged in terms of money or power, being able to afford expensive and often superior legal advice. This is known as lobbying (Hasen 2012). It may also come from the underprivileged: Revolts and revolutions are movements against the established legal order.

Petitioning has a long history, predating the modern state:

> 'In early modern Europe, petitioning was ubiquitous. It was often the only acceptable way to address local and national authorities when seeking redress, mercy or advancement. As a result, petitioning was a crucial mode of communication between the "rulers" and the "ruled."'
>
> (Royal Historical Society 2016)

In a historical case study about the Netherlands Janse found that 'it was not progressive reformers, but most prominently conservative Orthodox Protestants who organized [petitions]' (Janse 2019). Their main success was the challenge to a education reform bill brought by the liberal government in 1877. They mobilized nearly 800,000 signatures in short time and established a powerful movement, including other social groups, among them even Catholics.

With the availability of online petitions existing laws are challenged more frequently, across the globe, even in authoritarian societies (Detenber et al. 2014). Petitions have become so frequent that it has become difficult to keep track of them. Text mining technologies have been suggested to identify clusters of topics of petitions more easily (Suh et al. 2010).

Different countries deal differently with petitions. In the UK, for example, a petition needs to reach 100,000 signatures before it will be dealt with in Parliament. Parliament is sovereign to decide what to do with the content of the petition.

Ireland has seen a surprising change in its same sex marriage and abortion legislation (Murphy 2016), in which forms of deliberative democracy (citizen assemblies, referenda) played an important role (Farrell et al. 2019). Liberal groups and the government prevailed in a traditionally Catholic and traditional society. Interestingly, while the same-sex marriage bill was introduced by government in 2015 and confirmed in a referendum, conservative opponents launched an (unsuccessful) petition against the result.

This overview underlines the notion that law is exposed to citizen participation in various ways. The main influence is on the level of legislation. The judiciary itself tries to limit and channel these activities in specific ways, aiming to shield its operations from illegitimate interference. This pertains to the application of law, where courts and judges are holding ultimate decision power. Technological developments make this barrier more permeable than previously thought possible. With regard to the production of law the political constitutions provide opportunities for public participation, which are again legally circumscribed. In summary then, the law seems to exercise more power to look after itself compared to medicine and journalism.

AI applications in law

In this section, I will address the relation between AI and law, asking two different questions. The first asks how the law deals with the emergence of AI systems, coming back to an issue I raised in Chapter 2 where I discussed the contribution of Actor Network Theory and the problem it poses. The second question asks if AI systems are, or will be used in legal decisions. Is the robo-judge a possible reality?

The literature on Actor-Network Theory has made an important contribution in emphasizing the phenomenon of distributed agency. A weakness in this literature is the lack of discussion about the degree to which non-humans can be understood to 'act.' The sociologist Werner Rammert argues that there is no difference between humans and non-humans with regard to causal efficacy: many tasks can be executed by a human or a machine. Some types of machines even introduce contingency; that is to say they can choose from a set of options without humans interfering with the choice: autonomous systems, based on machine learning are a prime example. Where humans are different is with regard to intentionality and meaning creation. But even here machines could be programmed 'as if' they were pursuing defined goals.

Legal theorists have commented on the problem. Wagner uses the terms robots and autonomous systems interchangeably, underscoring the important characteristic of behavior 'determined by a computer code that allows at least some room for "decision–making" by the machine itself. In other words, a robot is not entirely under the control of human actors–particularly in accident situations. This feature distinguishes autonomous systems from more traditional devices that run on software-deterministic computer code that does not allow for machine learning or autonomous decisions by the machine' (Wagner 2019:591).

Teubner indicates that the legal system decides about the legal status of technological applications, including questions such as: does the application have intentionality? This is especially relevant for AI: 'The law which traditionally opened itself to real people and to collective actors now increasingly makes itself responsive to the new associations between humans and non-humans' (Teubner 2006:520).

Rammert pointed out that we should treat the question of agency as an empirical question, following a pragmatic line of reasoning. So what does empirical data tell us? Wagner asserts that as of 2019, robots and other autonomous systems are not recognized as persons or legal entities. 'This is true for both U.S. and E.U. jurisdictions. To date, no legal system classifies robots or other autonomous systems as legal subjects, instead leaving them to share the classification of other items made by humans: objects. Objects can be owned by legal subjects but cannot themselves be subjects or own objects. This means that the law attributes their actions–and omissions–not to the objects themselves but to the subjects responsible for them' (Wagner 2019:592).

Wagner reports about efforts by the European Parliament to address civil liability for damages caused by robots. It also envisioned, at least in the long term, the creation of a special legal status for robots: their recognition as electronic persons, or 'ePersons.' In legal reasoning this means to give robots the status of a subject, or legal person. As Wagner says, 'the criteria for personhood–intelligence, free will, consciousness, intentionality, and emotions–amount to little more than a laundry list of features that distinguish human beings from other creatures, and from objects. It remains unclear, however, which features are necessary conditions for the recognition of personhood and which are not' (595).

Wagner further observes that for this reason several authors have resorted to a gradual definition of personhood, echoing Rammert's suggestion from above: 'The fact that the properties that define human beings are a matter of degree and do not allow for "either/or" determinations has led scholars … to a gradual concept of legal personhood.'

However, this line of reasoning seems to lead into a conundrum: 'Personhood is "associated with the legal capacity to act," which in turn may involve the capacity to contract or the capacity to commit a wrong. While autonomous systems may be subjected to civil sanctions, they are not enough of a "person" to also subject them to criminal sanctions. Still, reference to the "legal capacity to act" remains obviously circular, as the capacity to act depends on the recognition as a legal subject' (Wagner 2019:595).[3]

Teubner ponders the question if the legal system is set up to defend human control against AI, to defend social order against a potential takeover of non-human agency: '[T]he personification of electronic agents amounts to a socialization of things which is troubling our time. In the dynamics of alienation and reappropriation, the question for the law is: will new constitutional guarantees be in a position to counteract the infamous "Code", the electronic architecture of the internet? Will economic, social, and technical transactions run by electronic agents be brought back under human control?' (Teubner 2006:521).

While robots do not (yet?) have the status of a legal person and ePersons seem to be implausible, there are existing IT applications in law itself which are of relevance here. Computational law, the automation of legal analysis and legal reasoning, through logic or data analysis, is reality. This leads me to the second question from above, the degree to which AI ('the robo-judge') poses a challenge to law.

A decisive difference in applications exists between AI systems which produce knowledge, which in turn is used as advice, which can be adopted or ignored by a human decision-maker, and AI systems which put the knowledge into practice without further mediation by a human decision-maker. This distinction parallels the distinction between human experts who advise decision-makers, and human specialists who implement a course of action.

Expert systems which provide information for decision-makers are posing no additional conceptual problems compared to human experts. This is different in the case of systems that implement a course of action without further human intervention. Le Sueur calls this automated decision-making and reckons it 'may achieve more consistent implementation of [laws] … than can be offered by human officials. Automation based on the application of objective criteria holds out the promise of legal certainty (like cases are treated identically), the elimination of bias, ensuring that no irrelevant considerations are taken into account, and that all relevant factors are included. To this extent, automation can be regarded as enhancing the rule of law' (Le Sueur 2016:8).

Let us consider the robot judge. As Bhorat (2017) notes robots are biased in similar ways as humans: He cites the example of a Microsoft AI Twitter chatbot which 'became racist, sexist, and anti-Semitic within 24 hours of interactive learning with its human audience.' While algorithms may simply replicate existing human biases, 'the distinguishing feature of AI over an algorithm is that it can behave in surprising and unintended ways as it "learns." Eradicating bias therefore becomes even more difficult, though not impossible. Any AI judging program would need to account for, and be tested for, these biases.'

Concerns about these developments led the European Union to include a clause in its General Data Protection Regulation proclaiming 'the right not to be subject to a decision based solely on automated processing.'

Nevertheless, AI systems are used in public administration where existing applications seem fairly uncontroversial, as Le Sueur points out: 'A future (still distant in 2015) in which all government decisions affecting citizens are taken by computers feels like a dehumanised dystopia. But some automation seems uncontroversial. We do not, for example, expect officials in HM Revenue and Customs to work out each taxpayer's liability using pen and paper to make mathematical calculations' (Le Sueur 2016:13).

Here AI systems are conceived as trivial machines which have a clearly defined link between input and output which enables them to execute a regulation or law. An example is the 'Employment Status Indicator' (ESI) tool provided by the UK tax authority. It allows businesses to find out whether a new worker will be an employee or a self-employed contractor. The tool provides an answer based on a computer app questionnaire (Le Sueur 2016). This system may be even better than humans when it comes to decision-making. It would replace human specialists who are professionally trained to perform routine tasks. Again, it seems problematic to call such systems AI as they do not have the element of learning. They would not be experts in the sense of dealing with uncertainty and value conflicts. This is where judges and lawyers come into play.

Making legal judgements is a complex activity which involves not only legal expertise but is connected to personal experience and commonly shared values. Judges are social beings inhabiting a lifeworld which gives meaning to the cases they adjudicate. It has been said that '[m]any of the central concepts in the judicial application of the law – such as "justice", "reasonable care", and "intent" – are deeply enmeshed in the fabric of human life' (Sartor and Branting 1998:105).

Another author concurs regarding the unique characteristics of judicial decision-making and the limits of AI: '… reducing judicial reasoning to simple logical models is untenable, … attempts to formalise judicial decision making are fundamentally incompatible with judicial function and the application of artificial intelligence to the law is bound to fail' (Hanna 2019:27). If this is the case we do not know, but currently AI systems are not capable of replicating social intelligence and 'common sense' (Frey and Osborne 2013). If they are able to master this bottleneck in the future, they might succeed. It is another matter if we are going to like this success.

As if echoing similar arguments in the debate about AI in journalism, there is concern about essential characteristics arguably missing from AI systems, especially emotional intelligence: '[N]o pattern-matching algorithm can accurately simulate judicial reasoning because the judiciary displays emotional intelligence, and is tasked with producing reasonable solutions in cases where the rules, facts, and how they fit together are controversial' (Hanna 2019:28). This was already the opinion of commentators in the late 1990s when Sartor and Branting (1998:106) acutely stated: 'No form of legal reasoning seems to depend more heavily on uniquely human abilities than the decision-making of a judge.'

Another challenge to AI is the fact that the law is not consistent, that there are different levels of authority, and that there is leeway in interpreting the law.

At least in the common law tradition, as Gava (2009:79) maintains, 'There can be no absolutely right answer to contested legal issues because reasonable practitioners of that method can and do vary in attempting to apply their understanding of a vast and unruly body of legal rules and principles to an essentially infinite set of fact situations.'

This has implications for the prospect of AI applications in legal practice, as Gava explains:

> Indeed, given the immensity of the legal materials it is unrealistic to expect judges to have a mastery of the law. There are just too many rules

> and doctrines with too many competing lines of authority (as well as inconsistencies) for the law to be reduced to the equivalent of an algorithm. Nevertheless, within these inescapable constraints the judges are expected, as far as is humanly possible, to be faithful to the common law tradition, and their reasoning and decision-making should not be understood as giving licence to freewheeling choice and innovation.
>
> (Gava 2009:79)

What Gava acknowledges is the fact that there might be more than one legitimate legal verdict, and that the judge selects one, with reasons that stand up to legal challenge. He is not comfortable with the idea that an impersonal algorithm could be given the same freedom of making such a choice, even in the knowledge that the decision might not stand up to scrutiny and could be challenged.

Different types of legal cases might lend themselves to different types of decision-making. Some think that AI might be better than human judges at using precedent to decide or predict cases. This could lead to a situation where AI systems decide cases 'based on the facts,' whereas human judges 'could be reserved for higher courts where more complex legal questions need to be examined' (Hanna 2019:35).

Echoing the practices in journalism and medicine, we can distinguish between routine, low-level activity which can be automated, and complex procedures where AI only plays a role in the creation of a database, but does not go any further. AI could be employed in a supportive function, without gaining the autonomy it has in journalism.

Narratives in the field of legal practice have again highlighted the issue of bias and impartiality, and of control and responsibility. The commentators who developed the narratives reviewed above are mainly legal and social science scholars. Law has more autonomy to reflect on the challenges to its profession. It recognizes a circumscribed role for citizen expertise and shows less presence of AI companies which try to sell their products to legal practitioners. This underlines the different status of law compared to medicine and journalism.

What do the examples of the professions tell us?

The three cases of journalism, medicine, and law have shown some things in common. In all cases expert systems have been spreading which provide information and decision tools. Non-human agents have become ubiquitous and often better than humans in accumulating, searching, and filtering data. However, it is still the prerogative of a human to make the final decision.

The situation is different where such systems directly implement a decision. However, the diffusion and penetration of robots is uneven in my three case studies. We can distinguish journalism on the one hand from medicine and law on the other. Journalism has seen the most extreme penetration by AI systems which not only have provided information tools for journalists to use, but in some cases AI systems have replaced the journalist as author. In the case of the law the 'robo-judge' has not emerged and it is doubtful it ever will. AI systems exist which for example predict the likelihood of criminal re-offending, information which is used by humans in the crime and justice system. Medicine has seen the introduction of AI for the purposes of more accurate and faster diagnostic tasks, with the potential that the diagnosis will be taken for granted by humans without further inspection. In this sense AI might go deeper than in law, but certainly not so deep as is the case with journalism.

What can account for the difference between these cases? Why are AI systems more widespread in journalism? It would be tempting to cite the power of professional groups for this difference. Journalists are the least professionalized group among the three under consideration here. There is not even a need to have professional credentials to become a journalist. Lawyers and doctors on the other hand are the paradigm professions with high barriers to entry, and a strong professional ethos. This corresponds to a high social status, quite in contrast to journalists.

But there is something else at stake which would explain the deeper penetration of AI into journalism, but not law. This has to do with the function law has in state administration and governance. Because judges serve the state, they have a very clear mandate which must not be put in jeopardy by other actors, be they human or non-human. Medicine and journalism are mainly private institutions and are therefore open to competition. In law this is the case with private lawyers, but not with judges and prosecutors. If this line of reasoning is correct, one could expect future AI intrusion into those professions where private competition prevails, and where it is less regulated. Of course, much depends on the legal framework and the clients' needs. There is always the prospect that private companies introduce technological innovation, absent legal restrictions. Such innovations could change the nature of the game, leading to a change in social and legal practices.

Another argument relates to the nature of the task environment. Legal scholars usually point to the unique characteristics of the legal practice which involves evaluation and discretion. We have seen that these tasks have been assigned to AI in other contexts, so the reason for their difference cannot be located in the different task environment. When a judge rolls the dice in denying or awarding someone a more lenient sentence, the same could be done by an AI system. However, when a judge bases her decision on prejudice and does not roll the dice, an AI system might be preferable, depending on what the nature of the prejudice is, provided the AI system has not been programmed using these, or other, prejudices.

Commentators believe that the very nature of the common law system defies the application of AI: 'These limitations and obstacles as posed by common law theorists are fundamental and go to the very nature of the common law itself. Accordingly, any attempts to apply AI to judicial reasoning in the context of common law would render the system something else entirely. These difficulties which cannot be overcome, at least for the foreseeable future, include the absence of uniform written expression of common law principles …' (Hanna 2019:42).

Arguably, it is within the scope of AI systems to produce such a written expression of common law principles. And the answer to the question of why this is not being proposed is given above: it would render the system something else entirely. Maybe the system would morph into a civil law system? But so far even civil law systems have proven immune against the AI intrusion where one might have expected a more clock-like execution of judgements following prescriptions in the legal texts.

As suggested above, this has to do with the logic of the legal system. Legal reasoning is messy and contested; the rules are made up and are changing in dialogue with society. Most importantly, in the end everyone is convinced that the decision that was taken is intelligible and can be reconstructed—and contested. AI systems would make the process perhaps less messy but unintelligible. Not only citizens, but judges and lawyers, too, would feel alienated from the verdicts of such a procedure.[4]

How pervasive will AI be? Some authors like Arvind Narayanan think there is too much hype in this field, with many snake oil sellers. Companies try to exploit the success of achievements like AlphaGo and engage in aggressive marketing to influence public opinion. There are bogus claims of being able to identify the best applicant for a job using AI (Raghavan et al. 2019). The US public has been led to believe that the automation of all jobs is only ten years away. We need to sort the hype from reality. There are indeed areas where we do see genuine, rapid technological progress, for example in content identification via music apps like Shazam, reverse image search, face recognition, medical diagnosis from scans, speech to text translation, or identification of deepfakes.

Even in celebrated cases there is reason to be cautious. Computer scientists point out that with regard to Google's self-driving cars, they are driving on virtual maps, not on real roads. The machine-learning component is restricted to an environment which is controlled. These cars would do badly out 'in the wild.' Madrigal (2014) says 'But the "map" goes beyond what any of us know as a map. "Really, [our maps] are any geographic information that we can tell the car in advance to make its job easier," explained Andrew Chatham, the Google self-driving car team's mapping lead … Rather than having to figure out what the world looks like and what it means from scratch every time we turn on the software, we tell it what the world is expected to look like when it is empty … And then the job of the software is to figure out how the world is different from that expectation. This makes the problem a lot simpler.'

Narayanan expresses some important concerns: 'The fundamental reason for progress is that there is no uncertainty or ambiguity in these tasks – given two images of faces, there's ground truth about whether or not they represent the same person. So, given enough data and compute, AI will learn the patterns that distinguish one face from another. … and that's exactly why we should worry' (Narayanan 2019). It is interesting that Narayanan thinks that there is progress in what he calls automated judgement, e.g. spam detection, detection of copyrighted material, automated essay grading, hate speech detection, or content recommendation. But he cautions that 'AI will never be perfect at these tasks because they involve judgment and reasonable people can disagree about the correct decision.' This is an important point which has implication for democratic decision-making (see the final chapter).

There is another area where no progress is to be expected. This relates to predicting social outcomes, such as criminal recidivism, job performance, terrorist risk, or predicting at-risk kids (Dressel and Farid 2018). AI is not performing better at these tasks than humans and there is a danger if we were to push towards AI. The transfer of power from domain experts and workers to unaccountable tech companies is a serious concern. This means de-professionalization goes hand in hand with privatization of data. AI works like a black box and therefore its results cannot be explained. But because it has the veneer of accuracy reliance on these systems may be an attractive option for policy-makers, and private consumers. In a repetition of the wonders 'the computer' or 'the internet' have brought us, it is only one more step in a similar direction, or so the critics claim.

In his book *Who's driving Innovation* Jack Stilgoe comments on this issue with regard to self-driving cars:

> When a self-driving car is involved in a crash, conventional ways of learning from car crashes will not be much help. It is hard to extract adequate explanations from machine learning systems. The creators of algorithms may not want to reveal trade secrets, they might not see the need for accountability, or they might justifiably claim that full interpretability of a deep neural network is impossible. The designers of neural networks are not clear, for example, how their algorithms know that a dog is a dog and a tree is a tree, although early research suggests that the machines and humans analyse images in very different ways, which opens up the possibility that a self- driving car's software could become confused either by mistake or malice and we wouldn't know why. The first fatal crashes involving software-driven cars have already revealed how complicated accountability will be.
>
> (Stilgoe 2020:46)

A final note on the overall prospect of AI for the future. Narratives have paid much attention to utopias and dystopias. Terms like transhumanism and singularity are the concepts that signify a radical break with society as we know it. This may or may not come to pass; it is difficult to predict the future. But we need to think about the mechanisms which would make such developments more likely, and those that would make them less likely.

AI is not a clearly defined package that we either embrace or reject. AI applications will become, and already are, part of our life where they allow specific groups and individuals to gain an advantage. This may be a benefit in terms of speed, of quality of information and advice, of reliability, and so on. As there is a cost associated with the development and purchase of AI systems it is more likely that their diffusion is associated with wealth and social status. Their introduction could be voluntary and piecemeal, exacerbating social inequalities. Their introduction could also be mandated, or quasi-mandated, through surveillance technologies driven by large corporations and state bureaucracies. Other scenarios are possible. What seems to matter here is the potential of civil society to push back against developments that are going against basic values and expectations in modern societies. If liberty and equality are still of concern in liberal democracies there will be limits to the brave new world.

Summary of Part II

In Part II, I examined the role of science for decision-making, using the cases of climate change and Covid-19. I also looked at the role of the professions and the challenges they face. My analysis has shown that science as such cannot be equated with expertise. The IPCC provides assessments of the state of scientific knowledge, but not in the sense of advice for decision-makers. Expertise is elicited in different national contexts which operate under different rules and traditions. Scientists are engaged in strategies of boundary work, but these are not always successful. Advisory panels in the policy-making process embody forms of expertise on the national level, which have been extensively researched and are fairly well understood. The IPCC presents a different type of expert body providing global assessments which operate to some extent detached from national governments. While governments nominate the authors for the report and approve the Summary for Policymakers document, the IPCC has no advisory function for governments.

Overall the IPCC exemplifies the role of a commentator-cum-science arbiter in public discourse. Climate scientists also act as advocates, invoking the IPCC as epistemological authority and bedrock of evidence. The IPCC also defines policy alternatives, if only implicitly and indirectly.

In the case of Covid-19 expert advisory bodies have been set up on a national level. Epidemiology has taken centre stage and provided the evidence and justification for government interventions. Compared to climate change, such interventions have been far reaching, affecting basic liberties and foundational principles of economic activity. Epidemiologists have developed the narrative to justify such measures. Their role goes far beyond the commentator as they have openly advocated specific measures. They also do the work of science arbitration, but this process is not performed publicly. In contrast to climate change, there are no lengthy scientific reports which summarize and evaluate the knowledge base.

I have also examined examples of professional expertise where we witness different characteristics, namely the roles of specialist, advisor, AI system, and commentator. Different fields of professional expertise show different blends, resulting in different hybrid constellations. Not only are there differences among different fields of expertise, but there are also national differences, too. One should be wary of generalizations given what we know about different national institutional contexts and traditions.

I will now take stock of the theoretical concepts presented in part 1 and discuss their usefulness in the light of the empirical examples from part II.

Automated decision-making has been a topic of research in areas of medicine and law for some time now. On a general level, one can say that such systems work in one of two ways. They are either implementing decisions without further human intervention. This is the case with robots. Or they do involve human intervention and the system is 'one tool among several for a human decision-maker who ultimately brings their judgement to make the final decision themselves' (Cobbe 2019:3).

For such system to operate, they need to be able to identify, collect, store, and sort information. Classification is a key activity in this regard. As Bowker and Star note in their account of classification and its consequences, 'each category valorizes some point of view and silences another,' and there is a long history of lives 'broken, twisted, and torqued by their encounters with classification systems' such as the race classification system of apartheid South Africa, or the categorization of tuberculosis patients (Bowker and Star 1999).

While proponents of AI systems have claimed that algorithms help us classify more 'objectively,' critics argue that this is unlikely as the algorithms are based on assumptions and decisions made by humans: 'human judgment [is] still involved in designing the algorithms, choices which become built-in. This human work includes defining features, pre-classifying training data, and adjusting thresholds and parameters' (Burrell 2016:3).

People sometimes expect that experts can predict the future. Even if this expectation is not expressed clearly, or its possibility even denied, experts make an assessment of what is known ('the facts') and provide advice about a course of action. This involves some understanding about connections between the past, the present, and the future. Causal models and assumptions are (a latent, or tacit) part of the expert's repertoire and skill set.

Experts are not only called upon as forecasters. Experts often need to know the past: Forensic experts tell in court what happened before, during, and after a killing, financial experts explain why the housing market crashed, military experts tell the government that a foreign power has developed a secret weapon. Close to the examples discussed in this book, climate scientists provide an account of past climates of the last centuries or millennia, doctors study a patient's family history of illnesses, and judges release a suspect on bail based on their previous life history. News media is full of comments about what happened and will happen next.

Nevertheless, the action-oriented, and future-oriented aspects of expertise are what catches the most attention. Will we experience a climate catastrophe? Will I develop the same heart problems as my father? Will the offender re-offend? If we only knew the answer to these questions, so we hope, we could take much better care and could be much better off. Unfortunately, in all three examples, the relevant expert will tell us that they cannot be sure. Probabilities are all we get.

Notes

1 It goes beyond the scope of this book to examine this topic. Suffice it to say that such neo-utopian attempts are seen as doomed by several commentators. See e.g. Schiffer (2018) and the cited literature, especially Gardiner (2012), who points to 'capitalism's ability to appropriate anticapitalist movements … contemporary capitalism is not simply a superstructure hovering above social life, existing only in the realm of the "global," but, rather, also infuses the rhythms of ordinary lives, structures society temporally and spatially, and even modifies the body … Contemporary political ecology scholars and eco-activists emphasizing the "local" as a politically charged arena for cultural change, with the potential to resist and critique globalizing capitalism, may find that capitalism's recuperative quality—its ability to absorb competing utopian visions into a commodified lifestyle—has beat them to the punch by operating at multiple levels, including what we may call the hyperlocal/ corporeal.'

2 The first newspapers were the *Relation*, which was published in Strasbourg in 1605, and the *Aviso*, released in Wolfenbuettel in 1609, see Fürst et al. (2015).

3 According to news reports, there are efforts underway in Estonia to grant robots legal status, 'purportedly somewhere between a "separate legal personality" akin to an incorporated entity, and an object that is someone's personal property akin to a possession. The Estonian economic ministry has termed this legal status akin to "robot agents", and commentators suggest that such legislation could "help determine who is responsible when AI-controlled vehicles or machinery are involved in an accident"' (Hanna 2019:25).

4 Nevertheless, there are examples where the prison sentence of defendants has been based on the verdict of an algorithm, and the appeal of the suspect has been rejected by the US supreme court, see Harvard Law Review (2017).

Part III

Chapter 10

Conclusion

Key institutions of modern society were built on monopolies of knowledge, sometimes ennobled as 'truth': the church, science, law, medicine, and later journalism. Simplifying slightly, one can say that these monopolies have been broken up over the past centuries in the case of religious doctrine, in the past decades with regard to science and medicine, and in the past years with regard to journalism.

In all cases, there has been an influx of knowledgeable outsiders into the domain of expertise. This trend has been accompanied by the rise of AI. The combined forces of citizen power and AI has the potential to further dissolve and reconfigure existing domains of expertise. This new constellation is a product of social and technical change, which in turn leads to further change. How will this look like? How can we understand it?

Whatever the answer to these specific questions, a preliminary hypothesis is that we will not witness the 'death of expertise.' Social actors will continue to seek advice from knowledgeable people and systems who they trust. In fact, the demand for expertise is ever growing in a world which is uncertain and produces ever more information about issues that have potentially serious implications.

However, the landscape of expertise is not uniform and so we should expect that the changes will not be uniform. Law seems to be more resisting to the double prong attack from citizens and machines, while journalism is in the process of profound change. Perhaps medicine will go down a similar path.

An early typology by Habermas still captures important political concerns about expertise through his tripartite categories of a decisionist, technocratic and pragmatic mode of governance. Many commentators would agree that the pragmatic mode is the most desirable, yet we often seem to oscillate between decisionism and technocracy. More democratic forms of decision-making still seem to be squeezed out, as was the case at the time of Habermas' writing.

On the level of national policymaking, we may see a resurgence of decisionism, ending an optimistic era during which the emergence of the knowledge

DOI: 10.4324/9781003320227-14

society has meant that political decision-making had to be justified via certified knowledge. There are signs that this tacit assumption could be superseded. As the case of populist governments shows, the legitimation of political decisions not necessarily invokes research or expertise. The skilful manipulation of public opinion through powerful rhetoric may, at least for a while, replace the use of knowledge—previously known as trusted expertise. In essence, this means a return to a constellation that was common before the advent of the knowledge society. Those who govern use symbols and narratives to justify their actions. While this has always been the case, we have witnessed situations in recent years where political leaders have not even pretended to justify their decisions with expert advice.

On the level of professional expertise, the double pressure from citizens and technology will lead to a reconfiguration of professional expertise. In some fields, expertise will assume more the role as commentator rather than advisor. Law appears to be a special case as it seems to be less affected by the process of erosion described above. It would still be a major institution of society in which 'truth' resides, which is not contested because the democratic institutions do not allow this to happen. *Law is a function of the state, not a playball of the markets.* However, recent developments in politics, especially the actions of populist politicians have started to erode this institution, too. Historically speaking, we have been there before. But these efforts come from high up, from powerful human decision-makers, not from lay pay publics or expert systems.

Science is also affected by these changes. As I have indicated, the ever-present strategies of scientists to use boundary mechanisms to assert their epistemic authority are directly and indirectly challenged by the rise of citizen expertise and AI. The former arises in a context of the knowledge society where citizens without scientific or professional accreditation participate in public discourse. The latter hints at a different concept of what it means to conduct science, as the big data enthusiasm wants to leave behind established ideas of theory and causality. Scientists are still trying to monopolize, extend and protect their domain, and expel interlopers. But the success of such boundary work cannot be taken for granted.

Modern societies are at a crossroads. This is a platitude, as is the statement that it is too early to tell how developments will play out in the end. The advance of AI could be a matter of decades, but maybe of years. The Trump presidency has shown a rhetoric of rejecting knowledge-based policies, and the Brexit campaign has shown that certified expertise was not taken seriously. Both Trump and the Brexiteers have shown little restraint when attacking the judiciary and established media outlets. The prospect cannot be ruled out that in the coming years decisionism will triumph over technocracy, mobilizing democratic sentiments. However, national contexts count and it is not clear to what extent these developments can be understood as universal trends or if they will be limited in space and time, and remain episodes.

All-too-human and more-than-human

In part II, I discussed the double pincer movement challenging the professions. This is a historically novel situation. On the one hand, citizen science is questioning established expertise, on the other hand, technological systems are offering a replacement service. If we apply this theoretical device to the analysis to climate change, some interesting points emerge. First of all, computer simulations have been present at the very start of modern climatology as a scientific endeavour. From the 1960s onwards, numerical models have been developed to project future states of the weather and of climate. In this sense, artificial intelligence (but not machine learning) has been part and parcel of the enterprise, and without it, we would not have the same discourse as we have today. As I have emphasized several times, this way of knowing in climate science constitutes its epistemic core. Anything that cannot be captured in parameters that feed into climate models is likely to be marginalized, neglected, or ignored. We saw the low salience of issues to do with policy and governance issues or with social dimensions more generally in IPCC reports, and we see the daily media reports about physical indicators describing and defining climate.

In a similar way, the Covid-19 pandemic has been framed through a model-based approach via epidemiology. Numbers of infections, hospitalizations, deaths, and their development over short time horizons have been the key to decision-making. Other social concerns are addressed as an afterthought, much like in the climate case. Climatology and epidemiology are presented as the sciences which describe 'harsh reality' which cannot be avoided, much like the law of gravity.

Of course, describing computer models in terms of artificial intelligence goes against the current understanding of the term AI. Machine learning is the process whereby computer code emulates human decisions based on a vast amount of data on which it can be trained. AI proponents develop new code that is not based on rule following but on 'deep learning.'

Machine learning has made its appearance in research on climate change and Covid-19. There are about 2,000 papers on the Web of Science that contain the terms 'climate change' AND 'machine learning,' more than half of them published in the last two years. A similar number of papers has been published on 'Covid' AND 'machine learning,' most of them have appeared in 2021.

Youyang Gu, a data scientist, and creator of Covid19-projections.com came to the field of modelling the pandemic without any prior knowledge of infectious diseases. He developed a machine learning model to predict the direction of the pandemic, as he told the website Future a16z (Gu 2021). 'In a way, the biggest advantage I had was not knowing anything. Starting with a "blank slate" approach was extremely valuable, as many prior conceptions of how a virus operates did not necessarily apply in the case of this novel coronavirus. And since I had no prior conceptions, I based my results only on what the data was telling me — this turned out to be the right approach more often than not.'

This enthusiasm for AI tools is widely shared. It raises the question of which role theory and causal models play in understanding and explaining the world, a problem which was raised in Chapter 9. Proponents of AI believe in the virtues of data-driven approaches, where relations between variables, measured in a probabilistic manner, are the most important information. For them, correlation beats causation. Recall the plea from Gigerenzer above that 'in order to make successful predictions, we need a good theory, reliable data, and a stable world.' The rift between epistemologies could not be wider.

Long lists have been compiled for various research questions relevant to the Corona pandemic using AI (Ahmed et al. 2022). Its usefulness for developing vaccines has been claimed (Waltz 2020). The main example was a DNA-based vaccine developed by Inovio Pharmaceuticals. While the company claims its vaccine 'went from bench to bedside in just 83 days,' it still has not been approved for use. It is telling that the successful vaccines to date were not found using AI tools.

In the field of climate change research AI applications investigate research areas such as earth system modelling, future climate scenarios, or climate impacts and extremes (Huntingford et al. 2019). It is not clear how much of this research has been used for sifting through the fast-growing corpus of research publications and how it is used in decision-making contexts. It is telling that climate sensitivity, one key parameter which has not been nailed down, is a prime example in such debates (Jones 2017; O'Gorman 2019).

The other prong of the pincer coming at the professions is citizen science. With the rise of the knowledge society, an increasing part of the population has gained higher levels of education and has developed an interest in technical issues that not so long ago used to be the preserve of the professions. The same has happened in climate science, where citizens have started to debate claims about climate change. Helped along by online technologies, several communities of bloggers have emerged that challenged established climate science. More recently, climate activists have started to do the same. Covid-19 will not be the last case in the list of examples.

In his book *A Vast Machine*, Paul Edwards tells the story of the emergence of technical infrastructures around computer models for the prediction of climate. In the last chapter, he takes a look at the developing blog activities by several outsiders to the scientific elite, especially *Climate Audit, SurfaceStations.org* and *Clear Climate Code* (of these, only Climate Audit is still running under its original name).

Edwards zooms into the so-called 'hockey-stick-controversy' where scientists like Michael Mann became the target of these citizen science activities. He has some sympathy with Mann and his co-authors' predicament, being drawn 'into a years-long morass of hearings, letters and public defenses of their data and methods.' But he also warns against a reaction which just closes the wagons and dismisses these citizens as outsiders, or even denounces them as skeptics or deniers: 'If you close your science up, excluding outsiders by refusing them access to data and methods, not only will you raise suspicions and open yourself to accusations of elitism; you also may miss real scientific benefits from unusual critiques and creative ideas "outside the box" of your field's traditions' (Edwards 2010:426).

But Edwards is not totally convinced we have seen such a benefit. While he thinks some of Climate Audit and Clear Climate Code's activities were beneficial (such as the detection of errors in the temperature data), the same cannot be said about SurfaceStations.org. Edwards believes because their findings have not been peer reviewed, posting their conclusions on a public website was 'highly questionable.' Edward concludes: 'The cases I have described here represent relatively responsible efforts by outsiders to participate in the scientific process. Many other climate blogs and citizen science projects display poorer judgment, ranging from misinformed good-faith opinions to rants, religious screeds, and deliberate disinformation. It is hard to divine where this new openness will lead' (Edwards 2010:426–27).

This leads us back to the musical repertoire from above. In the end, Edwards intones the tune about the difference between real and fake expertise and the possibility to establish the differences without difficulty. But terms like poor judgement and deliberate disinformation sound more like an attack on specific persons and their opinions, indicating that the world is already ordered in such a way where there can be no doubt about such categorizations. But this relies on a shared understanding of what counts as good judgement, something open to debate.

It seems that any such account would rely on the possibility to define real expertise, something philosophers have often tried to do but failed (at least in my opinion). Such essentialist definitions are either leading into an endless regress or into making strong assumptions (Turner 2019; Watson 2019). The approach taken here is different. It draws on previous research that conceptualized expertise through social mechanisms. Credibility contests are fought by social actors and groups of actors. Boundary work and rhetorical strategies have been emphasized in this context. While these approaches are useful, their empirical relevance needs to be explored. I have done this by looking at different facets of expertise and their social context. Markers of distinction are defined by social expectations, which means they will differ across different types of expertise. Specialists and professional experts will be expected to have the necessary vocational qualifications and experience. Advisors will be expected to recommend courses of action, given the uncertainties and value conflicts. Commentators are expected to provide an informed comment that enlightens us about the world through their opinions. If this book serves a purpose it should be that the reader is aware of these different roles and functions of expertise.

I want to conclude this chapter with some notes on the limitations of my theoretical framework.

Firstly, the concepts of advisor, specialist, commentator and technical systems are supposed to create some useful tools to think about expertise in society. I have deliberately de-emaphsied traditional approaches which put scientists and professional experts in the focus of attention. As my cases have shown, there are instances where professional specialists are important, and my point is that they should be acknowledged in their role of providing expertise even more. What the case studies have also shown, especially the cases of climate change and Covid-19, is the relevance of science arbitration. This notion, taken from Pielke's work, has not received the proper prominence as it perhaps should have. Much of the recent discussions on expertise are on the goods and ills of advocacy and honest brokering, but the important role of assessing the state of knowledge seems to have disappeared from our view.

In the case of climate change the assessment of the scientific knowledge base is the prime task of the IPCC. Yet, as critics have pointed out, this task is not performed in the best possible way. There is a lot of leeway for members of the IPCC Working Groups to select specific accounts of the state of knowledge. Co-ordinating lead authors are in an influential position in this regard. At this level advocacy and honest brokering make their way into the IPCC reports, through the back door as it were. The role of science arbitration needs to be taken much more seriously, given what is at stake in these debates, and especially given the high expectations directed at climate science.

Covid-19 illustrates the problem from another angle. Here, no institutionally orchestrated systematic assessments of the knowledge have been conducted. There is no reliable database on important aspects of the pandemic, such as infections or mortality. Many reasons account for this situation, most importantly perhaps the fast moving nature of the pandemic. But the fact remains that there is little in terms of established, reliable data. The same can be said about theories that explain these developments. As far as I can see no institutional efforts are underway to address this problem. I am certainly not suggesting we should create a replica of the IPCC, given my verdict about its function and performance. Of course, systematic reviews have been done, more will appear in the literature, and reviews of reviews will take place. This will be 'self-organized' research, which tends to fall within specific disciplines and sub-disciplines. This does not guarantee that interdisciplinary aspects will be recognized. This suggests some kind of institutional format is required that enhances the prospects of proper science assessment.

Having said this, the selection of panel members, of procedures and results will lead to debate and controversy, this is the nature of the Post-Truth society. But not recognizing the need for such assessments cannot be the answer. My conclusion of taking the role of the science arbiter more seriously does not put me in the choir of the *Hey Jude* singers. They would suggest that the results of science assessments give us a guide for action and we should therefore just 'listen to the science'. My suggestion deals with the problem that Covid-related science has been oversold as politics needed to find ways to deal with the pandemic and justify the actions. Research results have been cherry picked to do the job. Admitting to huge uncertainties in this process was not part of the plan, neither in the alarming nor in the reassuring camp.

Secondly, the notion of boundary work has not been examined in a very systematic way. I take from Gieryn's important contribution that more empirical cases should be studied, and the theory should not be taken as dogma. Gieryn is right to postulate that such boundary drawing strategies are to be expected, and are important to study. However, it is an empirical question if these can be regarded as successful. As I have pointed out, the proliferation of knowledge sources, mainly in the role of commentators, has led to a competition about epistemic authority and political influence. Even the IPCC has to assert its position in public discourse.

Thirdly, the selection of cases in this book is idiosyncratic. I have justified the two-policy related and the three profession-based cases through their obvious relevance, at least from where I am looking at the world. Readers may agree or disagree on what I call 'obvious relevance'. However, it is not clear if different cases, in different contexts, would suggest different conclusions to mine. In this sense the proposed framework is the invitation to start new research. It is seems especially important to widen the scope since most research has been conducted in the Global North with scant attention to the Global South (for exceptions, see Borie et al. 2019; Falzon 2021; Trentmann 2019). Even within the Global North there is disproportionate attention to the US and European countries, which leaves large parts of the globe under-researched.

Chapter 11

Afterword: Governance, Democracy, and Social Change

I started writing this book as the people of the United Kingdom were preparing to vote in the EU referendum. I wrote the final draft while the world was in the grip of a pandemic. Both events are highly relevant to the topic discussed in this book, although they have not been part of the initial plan.

In the same period the role of social media platforms, through their big data and programming capabilities, have been in the headlines for influencing public opinion. It appears that both forms of expertise are intricately linked to each other, in both cases, in uncanny ways.

Brexit has been described as a once-in-a-lifetime decision, a decision that would alter the nature and future of the UK, and the nature and future of the EU. In the run-up both sides of the debate laid out their arguments. The debate split the country, and the major parties. It split the left and the conservative camps, and it split the world of business.

Covid-19 has been a sudden, surprising unfolding of a new epidemic disease which has taken many surprising turns. The political response was different in different countries over time, and there was again disagreement in the political parties and business organizations, in communities, families and friendship groups.

Project fear

In the run-up to the Brexit referendum in 2016 supporters of the Leave campaign denounced statements by the UK government and financial institutions as fear mongering. All would be well after the UK had left the EU. Project fear was a polemical slogan and it soon became clear that supporters of Brexit admitted that the future might not be, at least in the interim, so rosy. Nevertheless, this demagogic abomination has perhaps an interesting kernel of truth. A few years earlier, during the Scottish independence referendum, the political elite employed a scare tactic of emphasizing the potential economic disadvantage of an independent Scotland. The independence campaign was doomed.

DOI: 10.4324/9781003320227-15

This intervention seems to follow a script whereby experts warn laypeople about the dangers they don't know about. It is more commonly used in any kind of expert/lay relation where the knowledge asymmetry enables the expert to extract something from the client. In this example it is the maintenance of power. As Berger and Luckmann noted long ago, 'the general population is intimidated by images of the physical doom that follows "going against doctor's advice"; it is persuaded not to do so by the pragmatic benefits of compliance, and by its own horror of illness and death.'

How much could be said with certainty about its prospects, if Britain left the EU, or if it stayed in? No one could have know the future, but some established facts about these two scenarios needed to be considered. There were acrimonious debates about 'the facts,' leading several commentators to say that the vote was not about facts, but about fears. Fears that had been instilled by the leaders of both sides. During the debate the issues boiled down to a few, mainly fears about national sovereignty, immigration, and economic well-being.

One would think that when discussing these big questions experts would be able to assemble the facts. And they duly tried, publishing papers, reports, interviews, and comments. But the debate was highly ideological, dominated by fundamental values that were not up for discussion.

On the face of it, the debate centred around economic arguments: will the UK fare better or worse if it leaves the EU? Economic experts assembled the facts, indicating the UK would be worse off. The Brexit campaign suggested that it might be better off, as it would make huge savings by no longer paying membership fees. The famous slogan of giving £350m a week to the National Health Service instead was perhaps a carefully designed trap (Cummings 2017). It focused the debate on a side issue, and the question of if the Brexiteers were lying.

What was left out of the debate was the question of which kind of society the UK wanted to be, and which fundamental values would be important to it. The Remain side assumed that a technocratic defence of the status quo would be enough to convince voters, because they first and foremost want their economic interests protected. Non-economic advantages of being a member of the EU did not appear in the debate. The Brexit campaigners on the other hand captured the public mood, which was based on strong opinions, and on fundamental values about national pride. The power of ideas triumphed over mundane economic interests.

Brexiteers managed to mobilize voters around the ideas and values of national sovereignty and an optimistic outlook for the future of Britain. They hired experts who communicated this strategy in effective (if sometimes illegal) ways. In contrast, the Remain campaign relied on economic arguments, assuming that businesses and workers would favour EU membership for this reason. The battle for ideas was lost before it started. Remainers mobilized the wrong kind of expertise: institutional heavyweights who provided the numbers and forecasts, from the Bank of England, the IMF, the World Bank, to the Institute for

Fiscal Studies. These technocratic organizations could not excite people about future projects. All they had to offer was more of the same. It did not help that they failed to predict the financial crisis.

The Brexit saga has inconvenient implications for social scientists. First of all, it casts severe doubts on rational choice models which assume that ultimately economic self-interest, and the structural power of big business determines policy outcomes (Bell and Hindmoor 2014). It also shakes the beliefs of those who generally are in favour of more public participation in decision-making. A long history of critical scholarship is evident in several academic fields, including Sociology, Science and Technology Studies (STS), and Political Science. It may be stronger in Anglophone countries than in continental European, or Asian countries. It is politically left leaning, whereas the new critique of expertise sprung from a populist sentiment which is resentful about the privileged middle-class, left-leaning critics.

Some social scientists therefore have discovered the merits of technocracy and came to the defence of the beleaguered experts. Social scientists, as everyone else, are in danger of having their academic judgement clouded by political preferences. The fact that citizen expertise from below has been on the rise does not necessarily mean that it is a good or bad thing; it all depends on where you see yourself in this game. The same note of caution is appropriate when looking at the power and function of established expertise.

Does this analysis apply to Covid-19, too? Without any doubt, fear of the disease was a strong motive in the debate, but also suspicion about the power of a technocratic state. Rejection of epidemiological narratives and evidence led others, formerly critical of such technocratic practices, to come to their defence.

What kind of society do we want? This question was not posed in the Brexit debate, but at least implicitly in the pandemic. Protecting the vulnerable has been a principle which was widely shared and provided a narrative to justify harsh control measures. But it was not universally accepted.

The principle of solidarity was opposed by the principle of individual liberty and responsibility. Such libertarian values can coexist with a strong sense of social belonging, e.g. extended families relying on traditions and distrust of state intervention (as is the case in UK cities, see Bear 2020), or rural communities not wanting their folklore disrupted by a state which is normally trusted (as may be the case in Switzerland, Daum 2020).

Protecting the vulnerable is a collective action problem. Allowing individuals taking personal risks is not a sensible strategy where this leads to more infections and deaths. Limits to individualized decisions have been based on this reason which was shared widely, but also fiercely contested. A zone of compromise seemed to open up with large scale vaccination programmes in many countries. Slogans appeared that wanted to base Covid policies on personal responsibility. The public health logic, and the importance of solidarity, retreated to the background, making space for the logic of individualism.

Governing in the 21st century

In this sense there is a tension between two fundamental values which people deem important in social life, and which governments have in mind when appealing to the public or when justifying decisions. But there is also tension between two rhetorical devices that are used to influence public opinion: raising fears, and giving reassurances.

Governments, most of the time, under most circumstances, tend to reassure the public. Official expertise provides the knowledge resources and communicative repertoire for this task. There may be always lingering doubt in the mind of the public when listening to such reassurances. But the doubts usually get pushed aside, and people want to believe that things will turn out to be OK. Governments' use of fear as a tool in political communication had been reserved for exceptional circumstances, notably preparing for armed conflict.

As some examples discussed in this book have shown, a rhetoric of fear has been deployed by different kinds of experts, from the technocratic elite, to commentators, and activists. Even governments have taken to this rhetoric. One is reminded of Carl Schmitt's statement 'Sovereign is he who decides on the exception,' which has overtones of justifying illegal acts. It is tempting for governments to cite 'exceptional situations' as justification for using a combined rhetoric of fear mongering and reassurance in order to shape public opinion.

Let me now turn to the other side of the coin of troubling expertise. The widespread adoption of AI systems has been analyzed by scholars, but its implications are far from clear. In the case of human experts many are ambivalent about their role and merit. This ambivalence is entirely justified. It would be unwise to wholeheartedly embrace or reject AI applications. I want to pick one aspect of this *problematique.*

My question is the same as in the Brexit example. Instead of taking experts and expertise for granted, we should ask what kind of knowledge we expect from it. To be sure, AI systems are already here and will spread. The problem is that they are owned and developed mainly by private companies who are operating within legal frameworks that do not provide adequate governance of these systems (leaving aside that what 'adequate' means is contested). Where they are state-owned, they tend to be unaccountable as well. This problem needs to be addressed to discuss the potential merits and flaws of an expansion of AI.

A few big tech companies (Microsoft, Facebook, Google, Amazon, Apple) provide services, globally, that are in effect public goods. They provide, and are based on big data, and operate machine learning codes. In the span of a few years, societies have come to rely on these companies for their functioning (Iosifidis and Andrews 2019). The companies have a monopoly position, and civil society, even nation states, are weak in comparison. One author coined the term iLeviathan to describe them (Prainsack 2019). Four decades of de-regulation of state-owned infrastructure (railways, water, telecoms) has paved the way for the unquestioning acceptance of the principle under which big tech operates.[1]

The same verdict applies to big Pharma which has pre-empted public decisions through private initiatives. It was their decision what kind of vaccine to develop, with governments in the role of consumer, free to purchase specific vaccines, provided they had the resources and inclination to do so in a timely manner. This situation perpetuates with the evolution of the virus. As ever newer variants appear, private companies develop response strategies via their decisions about which vaccine to develop. When asked about the role of big pharma in this process, the head of health emergencies at the WHO, Mike Ryan, said:

> The question is: Are they partners or masters in this process? It should be done together. It should be done with the best interest of the world's population at the center, with science and evidence driving that decision-making. Whether manufacturers want to go along with that or not, I'm not their keeper or master; neither is WHO (cited in Branswell 2022).

We know that public goods can be produced by private organizations; the question is if it is a wise move to leave strategic decisions with global impact with private institutions. Suppose the nature of the public good is uncontroversial (public health), as is the way to achieve it (vaccinations). In the case of the period leading up to the production of the first generation of vaccines the private monopoly role arguably could be seen as legitimate, even if the COVAX initiative was the preferred approach.

However, with the occurrence of new mutations existing vaccines became less effective—even people with immunization become infected, passed the virus on, and become ill. Vaccines for new variants might be less effective against existing variants. What should be done under these circumstances? It is not obvious what the best course of action would be, and it certainly is not wise to leave the decisions about these questions by default with the companies.

Mike Ryan cautions: 'It's very hard to come up with the right answer. And therefore, there has to be some collective wisdom driving this. It can't purely be the decision of one manufacturer to say, "Well, this is what we're going to make, and this is what you're going to buy"' (Branswell 2022). This means that in hindsight, putting the decision-making power with regard to the first-generation vaccines into the hands of big pharma was a problematic step in the first place. It did not appear as a decision but as a default event, eased by the panicked reaction of most governments.

Returning to the role of AI and its contribution to the pandemic it can be said that improving the performance of AI systems typically requires new datasets, and more data which is collected in obscure and partly illegal ways, and owned by unaccountable companies across the globe. Under this de facto arrangement of public expropriation, no sensible debate about the merits of such systems can take place.

This only highlights the need for public discussion and democratic procedures to devise government structures that are apt for dealing with the issues at hand. This will not be possible if technocrats are in the driving seat, excluding the voices of civil society. The above-quoted WHO official is sceptical as regards the role of AI in the management of the pandemic:

> If we're looking at pandemic preparedness for the future it's not all going to be about technology and the machines that go ping and surveillance systems and AI and all of this stuff. We need a social solution for future pandemics, much more than we need a technological solution. Because we're dealing with a fractured world, fractured communities. We're dealing with a breach and prolonged breakdown in trust.
>
> (Branswell 2022)

Wither expertise?

Experts are said to be under attack but we need to distinguish which roles are affected. Professional specialists are not, nor are science arbiters. Also commentators will keep their place—they are used to public debates which can be controversial and no deference to their status can be expected.

Experts who make judgements in the light of uncertainty and value conflicts are the ones who have been under attack. This attack coincides with a changed political climate in Western societies which has led to a polarization between left-liberal technocratic, and right-wing populist tendencies. The latter have claimed the mantle of democratic virtue accusing the expert advisors of over-reach. Academics and left-leaning middle-classes have defended the experts, thereby expressing a preference for technocratic modes of governing.

In this constellation the critique of expert advice, and of experts more generally, has been couched in terms of denialism. This is a very different constellation to the period after WWII when left-leaning academics spearheaded the counter-culture and criticized technocratic modes of governance. The historical sketch below may be useful in this context,

It starts with a quote by Erich Fromm, a theorist close to the Frankfurt School, written in the 1970s. You may be forgiven if you thought it was from someone on the political right who warns about 'fake news' from mainstream media:

> Another helpful attitude is one of deep distrust. Since most of what we hear is either plainly untrue, or half true and half distorted, and since most of what we read in the newspapers is distorted interpretations served as facts, it is by far the best plan to start out with radical skepticism and the assumption that most of what one hears is likely to be a lie or a distortion.
>
> (Fromm 1992:96)

Writing in the late 1960s, Theordore Roszak, a theorist of the 'counter-culture,' made some very interesting observations about the role of technocracy in post-war American society. He wrote:

> In the technocracy, nothing is any longer small or simple or readily apparent to the non-technical man. Instead, the scale and intricacy of all human activities–political, economic, cultural – transcends the competence of the amateurish citizen and inexorably demanded the attention of specially trained experts.
>
> (Roszak 1968:7)

This, by now, familiar dichotomy between experts and laypeople appears to be a theme still alive today and is the basis of much recent writing on the subject. The ubiquitous nature of technocracy is nicely expressed in this quote: 'It is characteristic of the technocracy to render itself ideologically invisible. Its assumptions about reality and its values become as unobtrusively pervasive as the air we breathe' (Roszak 1968:7–8). Roszak sees the relation between technocracy and expertise as follows:

> The technocracy ... is a regime of experts, or of those who can employ the experts. Among its key institutions we find the think tank ... that seeks to anticipate and integrate ... everything on the scene. Thus, even before the public has become fully aware of new developments, the technocracy has doped them out and laid its plans for adopting or rejecting, promoting or disparaging... The experts who count are the certified experts. And the certified experts belong to headquarters. (Roszak 1968:6–7).[2]

In a similar fashion, Ivan Illich criticized that

> [e]lite professional groups have come to exert a 'radical monopoly' on such basic human activities as health, agriculture, home-building, and learning, leading to a 'war on subsistence' that robs peasant societies of their vital skills and know-how. The result of much economic development is very often not human flourishing but 'modernized poverty', dependency, and an out-of-control system in which the humans become worn-down mechanical parts.
>
> (Illich 1973)

And EF Schumacher famously said, '[i]n one way or another everybody will have to take sides in this great conflict. To "leave it to the experts" means to side with the people of the forward stampede. It is widely accepted that politics is too important a matter to be left to experts' (Schumacher 1971:130).

This critical attitude towards experts in the 1960s preceded debates about environmental and health risks, as documented and described by Jasanoff

(1990), Renn (1999), and others. These risk debates about nuclear power and chemicals in our daily lives rose to prominence in the 1970s—after being kick-started by Rachel Carson's seminal book *Silent Spring*, published in 1962—and were focused on the way the government reached decisions with the help of qualified experts (for an early insightful comment, see Weinberg 1972). The typical mode of addressing such issues in the USA was through an adversarial process of open debate where expert witnesses were cross-examined by the two main political parties, and by interest groups. This mix of science, law, and politics describes the process of public policymaking in the US from the 1970s onwards.

Roszak's observations concern the decade before this emergence of risk debates. He echoes the sentiment captured by de Tocqueville who wrote in 1835:

> '... if I seek among those characteristics [of the Americans habits, RG] the principal one which includes almost all the rest, I discover that, in most operations of the mind, each American appeals only to the individual effort of his own understanding'
>
> (Tocqueville 1835:489).

This appeal to individual understanding is what unites different tendencies that are critical of official expertise. The rationale for the critique of technocracy is that it diminishes or suppresses laypersons' concerns. This critique became an important element of the counterculture in the United States but initially less so in continental Europe, where hopes of a scientific technical intelligentsia were prevalent in the 1960s and 1970s. The so-called Richta report, published under the direction of Czech philosopher Radovan Richta in 1969 (Richta 1969), was heavily discussed in Germany where it became a reference point in debates (see, e.g. Böhme and Stehr 1986). It also became a reference point for authors in the USA who conceptualized the nature of the information age or post-industrial society (e.g. Bell 1973).[3]

To be sure, this optimism regarding the role of a scientific-technical intelligentsia resonated mostly with left-leaning academics. Conversely, authors from the political right have celebrated Anglo-American individualism, both from within academia (Lipset 2001) and from without (Davidson and Rees-Mogg 1997). Europe's social and political context was (and still is) different compared to the USA. No adversarial procedure was in place to deconstruct scientific claims in public, not in the Socialist bloc, and not in Western European countries. Policy advice in Germany, for example, developed along very different lines. Its corporatist arrangements express a desire for consensual policy development between interest groups and allow expert bodies to play a privileged role (Busch 2009; Campbell and Pedersen 2015; Hoppe and Wesselink 2014; Jasanoff 2005).

From the 1980s onwards, a big debate emerged about the role of expertise when the European Community morphed into the European Union. This process entailed two central visions, an inter-governmental and a supranational vision. An essential question arose with regard to the legitimacy of the rise of a European regulatory state which was seen as problematic in the absence of democratic legitimation (Gonçalves 1995; Majone 2005; Grundmann 1999; Rittberger 2005).

With the end of the post-war period, the end of the cold war, and global economic and environmental problems, the taken-for-granted role of technocratic bodies has been questioned. As uncertainty becomes emphasized, the prerogative of technocracy to deal with such problems is increasingly questioned (Corfee-Morlot et al. 2007; Turchetti and Roberts 2015).

The rise of the counterculture was a precursor to the current scepticism towards official expertise. Starting in the Anglo-American context, the critique of unaccountable experts and political elites eventually spread to continental Europe. Critique of experts was levelled at unelected 'Eurocrats' as immigration and issues of national sovereignty and identity arguably were more important than the issue of illegitimate power of experts. The latter became a topic in the UK Brexit debate, where we saw a cultural resonance with ideas about the 'sovereign individual' and the notion of 'common sense.'

Continental European discourse on the other hand had its own tradition which served as inspiration and provided cultural resonance. It ranged from the political right (Heidegger) to the political left (Frankfurt School) and offered a reservoir of 'cultural criticism' of modernity (Western rationality), or expert-based capitalism. Tellingly, Habermas' early advocacy of pragmatism, or what later came to be known as 'deliberative democracy,' was influenced by a dose of Anglo-American thinking, especially Peirce and Dewey's.[4]

Seen in this perspective, it can be argued that the recent 'populist backlash against experts' has important precursors in popular culture and philosophy of the post-war period. Of course there is a big shift in the political orientation of this critique as expert criticism and Euroscepticism became main talking points of the political right (Pirro et al. 2018; Szöcsik and Polyakova 2018). This is a far cry from political affinities of left liberals, such as postmodern theorists who pronounced the 'end of certainty,' suggesting an 'anything goes.' Their deep mistrust of grand narratives provided by dominant discourses was shared by some scholars in STS, who aimed at dethroning official expertise and its unquestionable authority (see Chapter 2).

Let me now turn to the social sciences where a sentiment has been growing to re-affirm the rationality and power of established expertise against a perceived 'attack on science' or 'truth.' I have used the musical metaphor of a simple riff to characterize this sentiment. Rather than demonstrating the rationality or power of expertise, such attempts are often politically motivated. If only the 'right experts' had the power to impose their insights we would be better off, or so the reasoning goes. There is a political divide which separates these different notions of expertise, and the tendency to abdicate previously held beliefs in order to 'prevent the worst' are noteworthy. Arguably this can be attributed to the fact that right wing political parties have used an anti-expert (or anti-elite) framing as a mobilizing tool.

The practices of Donald Trump suggest that during his presidency he has cut back the technocratic elements of his political reign, adopting a decisionist style of government in which 'political action cannot rationally justify its own premises. Instead, a decision is made between competing value orders and convictions, which escape compelling arguments and remain inaccessible to cogent discussion' (Habermas 1971:63). Similar developments can be seen in the UK, where government did not seem to want to base its transition to Brexit on advice from established expertise. But it would be far-fetched to interpret this as an instance of a pragmatic mode in which a constant dialogue with the public is taking place, as Habermas has defined it. Under current arrangements, elections and referenda have the function of 'spectacles and acclamation' (Habermas), as they do not establish the guidelines for future decisions.

Summarizing the change in public esteem of experts, Camille Limoges aptly observed:

> There was a time when the mobilization of experts was a taken-for-granted, unproblematic aspect of decision-making processes. Experts would 'educate the public' and, in so doing, prevent the eruption of controversies. It is upon such a basis that technology assessment, *qua* expert knowledge, was first established. Confidence in the power of expertise has now vanished.
>
> (Limoges 1993:417)

This critique was levelled at experts patronizing the publics, echoing criticisms by Roszak and others cited above. More democracy was seen as the answer. Part of this sentiment has been carried over into recent populist backlash. Within STS there is no agreement on how to interpret this new development; it remains a challenge. Broadly speaking, there are two responses. One response sees the critique of expertise as excessive, as documented in Bruno Latour's famous *mea culpa* (Latour 2004). In similar terms, Harry Collins thinks that some of the STS work is wrongly inspired by romantic notions of 'democratising expertise,' and what is needed now is the defence of expertise against the emergence of 'post-truth' (Collins et al. 2017). These two protagonists from STS have fallen for the alluring riff about the trustworthiness of official expertise. The coda about questionable motives of dissenting experts is missing or underdeveloped, but nevertheless, the basic hymn sheet is the same.

The other response is to vindicate the critique of expertise and to provide a more nuanced account of how lay perspectives should combine with official expertise in the process of decision-making (Irwin 2017; Jasanoff and Simmet 2017). This response comes from the dissonant choir of voices which problematize notions and practices of expertise, without developing a simple riff or tune.

Neither STS inspired critique of expertise, nor the critique inspired by the 'new populism' advocated a replacement of technical specialists. Rather, the concern is about the hidden political agendas that come in the guise of 'neutral expertise.'

The major difference between these expert critical voices is the political values that underpin their critique. In such situations of uncertainty and value conflict experts need to be challenged to lay open different options for action, the principles and reasons which underlie them, and the potential impacts of such actions.

Social scientists, and especially those working in STS, have long argued that facts and expertise are socially constructed. Just when this claim seems to enter the mainstream some of these same social scientists have become uncomfortable (see Fuller 2017 making a similar point). We should remind ourselves that politicians have always known how facts are made. They are not interested in facts as such ('best evidence'), but in power.

As John Gray put it in *The Guardian* (20 May 2017): 'Facts have been irrelevant in politics for a while. Why are liberals so shocked by fake promises?' In a similar way Jasanoff, writing about the need of STS to reflect its position towards power and expertise, wrote that 'self-examination reveals a more complex story of how we got here – and it was not from a past where politics was governed by pure *veritas*' (Jasanoff and Simmet 2017:752). The perplexity of some STS reactions can be attributed in some measure to the political polarization we are witnessing. But I would add to this two other aspects. The first is the lack of analytical distinction between different types of knowledge and expertise and their function in contemporary societies, a task I have addressed above. The second is the ignorance vis-à-vis different problems that modern societies are facing, a topic I will discuss now.

Towards a heterarchical world society?

One year before his death in 1998 German sociologist Niklas Luhmann published his *opus magnum*, *Die Gesellschaft der Gesellschaft* (published as *Theory of Society* in 2020). Reflecting on the evolution of distribution media, from writing to printing to electronic media, he argued that two trends stand out. One is that social order moves from the organizing principle of hierarchy to heterarchy; the other is that society is no longer spatially integrated (via nation states). Both trends go together. As a result we witness the emergence of world society in which attempts at hierarchical control and censorship are bound to fail, and in which nation states matter much less than before.

Today we may find this diagnosis puzzling, given the seemingly opposite trends in recent years. The return of nationalist politics, the emphasis on spatial separation during the Covid-19 pandemic, and state censorship of mass media in many countries speak to the contrary. Nevertheless, Luhmann clearly saw that society at the turn of the 21st century was going to be different to the past. His argument is worth reviewing.

His first comment refers to censorship and the rise of public opinion. Since the 18th-century authorities in Europe failed to establish censorship of the press effectively. He writes:

> 'Their failure, inevitable with the multiplicity of printing centers in various territories and with the rapidly increasing complexity of printed communication, finally obliged all hierarchies, including those of politics and the law, to come to terms with a fundamentally heterarchically communicating society. Since the eighteenth century, this state of affairs has been celebrated as the primacy of "public opinion"' (Luhmann 2020:187).

The current brouhaha about the dangers of social media, and the need to apply some kind of censorship, can be read in the same way. Established knowledge hierarchies resent the rise of yet another contender, and seem to be condemned to repeat the 300 year old experience.

Luhmann also saw the implications of what he calls 'computer technology' and how this influences the relation between official expertise and lay knowledge:

> '[Computer technology] attacks the authority of the experts. In principle, everyone will in future be able to check the statements of experts such as physicians and lawyers on his own computer. They may very well claim there is no scientific evidence for the efficacy of certain medications— and we find it nonetheless. Or that there are no court rulings on certain legal issues— and we find them nonetheless. Although it is difficult to check how knowledge finds its way into the computer, it can at any rate not be turned into authority' (Luhmann 2020:187–188).

This does not mean that 'the computers' are taking over. They do not establish a new authority but are part of a heterarchically organized society. The multitude of voices and claims makes it difficult to distinguish true from false. He asks: How can we be sure that claims are true? Have they been selected, stylized, falsified, or even invented? Are the experts in question activists and led by a 'special sense of mission' (*Sendungsbewußtsein* in the German original, R.G.)?

Luhmann's answer to these questions is intriguing. First, regarding the question of veracity of statements he recommends to apply the principles of second-order cybernetics, that is, to observe the observations, including one's own. There is no reality beyond this operation; all we can do is establish the credibility of claims through such recursive processes. The outcome of these operations leads to what we might call true description of reality. In this process decisions are made about the credibility of the sources. Can we believe the claims, and the claims makers?

This is where trust comes into play. Trust is an essential ingredient of all social life, but especially in a situation where different knowledge sources are competing. Trust has always been an important social mechanism, long before the age of electronic data processing. But now, Luhmann believes, things have changed. Trust can no longer be attributed to persons but becomes 'trust in systems.'

However, the world society is fairly uneven in this regard. There are high-trust and low-trust countries (Fukuyama 1995). The importance of boundaries which are spatial and political seems to be vindicated. Equally important is the trust in credible sources, which still can be linked to social status, and become personalized. The reputations of scientists and experts are often-quoted examples.

Luhmann disregarded national differences, preferring a view which posits world society as central concept: a world society that is functionally differentiated. Accordingly, he relativized personal trust and prioritized trust in expert systems.

I would concur that we have seen a move from hierarchical to heterarchical societies. While the spatial integration may have lost its grip during a phase of rapid globalization after the 1990s, more recently this trend has been countered by national and nationalist policies and strategies. That Luhmann could not anticipate this development is probably due to the fact that he was wedded to a conceptual scheme based on his previous work, which had examined the evolution of society. He introduced three dimensions of meaning-based integration in society: spatial, social, and temporal. He was convinced that the spatial dimension was losing ground in the course of history, and that the temporal dimension became more and more important. Societal integration based on physical presence, and physical boundaries, appeared to him as an anachronism.[5]

Nevertheless, the observation that trust is moving from personal attributes to expert systems is a finding supported by the case studies discussed in this book, at least to some extent. The growing importance of AI also testifies to this, but as we have seen, it is uneven in the fields of medicine, law, and journalism.

Inspired by Luhmann, my conclusion is that governing in this world means to be on top of public opinion and be able to influence it. The instruments of doing so are provided by techniques of persuasion. These rely on appeals to deeply held beliefs and emotions, combined with relevant information and their symbolic representation. These elements are combined in narratives for public consumption that rely on the double logic of arousing fear and reassuring the public. It seems we have moved far away from the notion of a rational society, or a rational mode of decision-making, based on best available evidence. It still remains as a powerful, if mythical measure for normative considerations and critique.

Voices from civil society can interrogate these narratives, bearing in mind that there can be no view from nowhere. All comments in public debates are tied to some kind of interest, and tied to a selection of useful information.

There will always be a technocratic element in the governance of modern societies. The nature of power, and of political systems means that advisors and lay experts will be co-opted into systems of decision-making and political communication. In this way the system keeps going, reproduces itself, and changes at the same time. It is the role of civil society, of commentators, to challenge the official narratives, and to devise new political projects, in more or less radical ways. As long as this process is democratically organized, the potential for learning and renewal exists. Expertise in its various guises is at the centre of attempts for finding acceptable solutions to pressing problems.

What about democracy?

Where do these developments lead us in terms of democracy? Do they undermine or enhance the prospects for democratic decision-making? Posing these questions requires a historical and institutional reference point. In my case this is the early 21st century, location is the Global North. The topics discussed in this book are written from this perspective, and I am aware of the fact that much of what I say is not applicable directly to societies from the Global South. I am discussing the problem of expertise and democracy in the context of a tradition of liberal democracy, with functioning state institutions and an expectation of broad parts of the electorate that these be maintained and protected.

My argument does not primarily relate to discussions about the 'hollowing out of democracy,' or democratic backsliding (Bermeo 2016; Runciman 2019). I am interested in the changing structure of modern society as a consequence of the changing nature of expertise and society.

What I have called 'the rise of the commentator' means that a wide range of citizens, groups, and spokespersons for organizations are participating in public debates, and all could be in the court of public opinion. Some might position themselves as 'authoritative,' and thus of a higher standing compared to 'ordinary' citizen experts, but the deference requested may not be forthcoming. The protagonists are rarely on a level playing field, and incumbent actors can be expected do everything to ennoble their own expertise, and to denigrate the challengers. The boundary drawing strategies are not only deployed by scientists but by other commentators in public debates as well.

However, treating dissenters, and citizens as ignorant, irrational, or immature is a suspicious move and often seems only one step away from denying them fundamental rights. Extended periods of suspending basic liberties in the Covid-19 crisis are a problem for democracies (Engler et al. 2021), as are calls to sidestep democratic mechanisms when dealing with the challenges posed by climate change (Stehr 2015).

Contrary to the early 20th century, when Max Weber developed his theory of bureaucracy, democracies have come to see the usefulness and virtue of public participation in decision-making about issues that were regarded 'technical' or 'esoteric' back then. However, there are limits to inclusivity, and at any point in time there are regulations in place that have been developed and implemented without much public debate. This is a necessary feature of complex societies which are based on a division of labour, to use an outmoded term.

Public administrations work well where problems are well-defined; this is their routine way of operating, as described by Latour's sewage plants at the beginning of the book. However, controversies can arise about these, and the very definition of well-defined and ill-defined, or tame and wicked problems becomes a matter of dispute. In other words, the jurisdiction of bureaucrats and established rules and regulations becomes contested. Is it undemocratic to keep the outsiders away from such debates? No doubt. Is it undemocratic to keep outsiders away from decision-making about these questions? It depends. It depends on the institutional settings under which such issues are processed, and if these are seen as legitimate. Some countries have referenda where such contestation is settled, even if only temporary. Other countries don't process these issues through party competition. But if the groundswell of public opinion indicates that voters are sensitive about an issue, and a desire for 'opening up' exists (Stirling 2008), then politicians will position themselves accordingly, and the issue might even become the central topic of an election.

In contemporary society, the principle of citizen participation has been gaining traction. However, there is a danger that more educated, and more vocal groups assert their interests, so they gain an advantage. This raises questions of social justice. We need to recognize that there is not one 'public,' but many 'publics.' The privilege of the educated middle classes is often resented, and resentment is prone to fuel populist politics.

While the multitude of voices needs to be recognized, decisions have to be taken. These decisions are political, not scientific. Democracies foster the belief that decisions can be reversed. The peaceful transition of power from government to opposition is the basic mechanism to achieve this. Party competition, separation of powers, and civil liberties are part and parcel of it. If a government fails to convince voters it will be the opposition's turn to do better after elections. This 'minimalist' version of democracy (Przeworski 1999) is convincing and appealing in many ways but still offers little insight into the fact that some decisions have lasting consequences which cannot be reversed. Of course, this verdict applies also to more comprehensive theories of democracy. But maybe it is not about reversibility in an absolute sense. Learning is what really matters. Democracies enable learning, from success and failure, not only from one's own past, but from other times and places too.

The rise of AI

A major challenge for every theory of democracy is the diffusion of technologies called Artificial Intelligence. In many decision contexts these systems have proven helpful, and their attraction seems irresistible. True, much of the claimed capabilities of such systems remain hype, at least for the time being. The question is, what if? What if they are developed into functioning devices that one day will deliver what is promised today?

From the viewpoint of democratic theory this would pose no big problem as long as these systems are implementing decisions for well-defined problems where we all agree on the aim that we want to achieve. Face-recognition or speech recognition have become ubiquitous examples. However, at times these systems fail and we need a backup that works, and works in a timely fashion. We still are able to key in the password on our mobile phone while wearing a face-mask.

There are more problematic examples. Decisions in legal and medical practice are often based on predictive algorithms. These need to be explicable and made transparent. Algorithms process information on the basis of different items to which weightings are applied, and both need to be understood and communicated. For example, calculating the risk for having a stroke can be done on several websites dedicated to cardiology. The items and their weightings are simple and transparent.

Many AI systems do not show the items and their weighting as they remain intellectual property rights of private firms. It seems obvious that no one wants AI systems that have the capacity to implement potentially disastrous courses of action directly. In analogy to the role of the (human) specialist, we require information on how rules are applied to cases. Where we are satisfied that this is the case, and we are dealing with 'routine' cases, there is an uncontroversial case for AI, too.

Things are different when we deal with situations of uncertainty and value conflicts. No degree of computing power will solve those situations for us. Reasons for decisions need to be discussed in public, and be exposed to critical scrutiny. In this sense a technocratic solution to the many social problems should not be an option. The paradox is that the proposal would merit a hearing, and we cannot pre-empt the outcome. As Runciman (2019:215) pithily stated, 'This is likely to be democracy's fate: the Trumps will come and go; the Zuckerbergs will keep going.' AI systems are developed by global information companies which operate on the basis of intellectual property rights, entrepreneurial freedoms, and a host of secret arrangements underneath the surface of their terms and conditions, and under the bonnet of their algorithm that harvest data and target individuals.

As I have argued, democracies need to reign in the power of these companies, an age-old question familiar to students of capitalism and corporate power. In this sense the problem is not new; it is about redefining the balance of power between public and private bodies. The difference is that we are dealing with tech companies who have provided many of the platforms on which citizen expertise is assembled. Participating in online exchange and debate has become ubiquitous and even mandatory, but it makes all participants into subjects of surveillance capitalism (Snowden 2019; Zuboff 2019).

A reconfiguration of knowledge regimes?

I now return to the discussion in chapter 5 on knowledge regimes. Knowledge regimes have been studied mainly with regard to economic policymaking. This research found that the interplay between research organizations and the state was reconfigured in a time of crisis, leading to more privatized, de-centralized, and competitive arrangements. In the case of climate change no such reconfiguration is apparent, and no attempts seem to be made. Maybe everyone is happy with the arrangements in place. While climate activists and some high-profile people speak of a climate crisis, most governments tend to pay lip service to this slogan. After all, climate change does not seem to be a crisis of knowledge, but of action.

With Covid-19 things could turn out differently. The process of gathering knowledge and mobilizing it for political decision-making is organized along traditional lines in each country, who all have their specific knowledge regimes regarding infectious diseases.

Covid-19 is perhaps the first instance where the knowledge creation and mobilization occur in real time, on a global level, in front of our eyes. Covid policies are affecting all of us. However, while we can witness the outcomes of decisions, and the reasons given for their legitimation, we as citizens are in the dark as regards the process behind the scenes. We do not know how the many disparate and conflicting studies and findings are selected, summarized, and interpreted, and by whom. We may know the names of the committees that advise governments, and perhaps their members, or of some celebrity scientists. We see some of them making public statements. We do not know how politicians make use of such knowledge and which advice they elicit. In times of increasing levels of public education and interest in scientific and technological issues this is an anachronism, even a paradox. It finds its expression in counter-publics that circulate and discuss alternative interpretations and recommendations.

Official expertise under the hegemony of epidemiology has not been open to citizen expertise. Contrary to other examples from the area of health we find the system largely closed to outsiders, and disagreement with official expertise denounced, even demonized. This particular knowledge regime seems to undergo a crisis with no signs of reform. Knowledge regimes for the 21st century need to be fit for purpose. This means they need to reflect the social reality of the knowledge society and provide arrangements for the inclusion of expertise from below. Professional voices and citizen experts need to be involved in the problem definition, and possible ways of addressing it.

A broad range of expertise is required, from advisors, professionals, commentators, and civil society. This will help to develop and support social practices that are embedded in our daily lives. We already have become frustrated by the on-off logic of large lockdowns that are based on territories and industries. Engineers may develop standards for well-ventilated rooms in which it is safe to congregate for prolonged periods of time. Community leaders may identify groups at risk and workable interventions. Policymakers need to find solutions to the competing jurisdictions between private and public bodies, and between different levels of government. They also need to incentivize efforts to find effective treatments. Health professionals may develop interventions that are quick and reliable and allow targeted responses. Global institutions need to ensure that resources to fight the pandemic are distributed fairly among countries. Pressure groups could campaign against the rent-seeking behaviour of big pharma. In sum, the governance of Covid-19 is something that needs to be established as a problem. Only when it is recognized as such will we be able to get away from an important but overly narrow definition of the problem, and open new avenues for intervention.

Technical expertise, as it has been traditionally embodied in modern organizations, has been taken for granted. They form the basis for routine operations. Whenever big questions were being asked, especially big political questions, this proved insufficient, and decisions had to be informed by more comprehensive epistemic resources. Scientific advisory bodies proliferated, which in a way replicated the logic of technical expertise in that they wanted to be shielded from public inspection and debate. This could have led to technocracy if there was not the political prerogative of making decisions, not only putting the advisors into place but also using their advice in a way not necessarily as intended by the advisors. Many commentators see this as the real problem of our times, urging politicians to 'listen to the science.'

My conclusion is different. To repeat: the rise of citizen expertise has led to the articulation of many concerns from within civil society which are based on knowledge claims. This expertise 'from below' has led to structural changes in democratic societies. It is no longer enough to refer to official expertise and to keep the outsiders out. Defining issues as well-defined in the field of scientific or professional expertise becomes a contested move. The functioning of technical systems, their construction, and their effects on society and the natural environment becomes a matter of debate. The many cognitive uncertainties and value conflicts create the opportunity for controversies in which potentially everyone has a say. The emergence of the knowledge society has led to more sources of knowledge, and its contestation, simply because there are many more commentators. The authority of official expertise is eroded. It sees itself under siege, and appeals to its higher epistemic competence, and authority. But in democratic societies such appeals are not always listened to, and it becomes difficult to discern who is right or wrong on a myriad of complicated issues.

This situation is here to stay, and it is futile to celebrate or condemn it. It creates more opportunities for citizen intervention and thus democracy, but it also creates uneven visibility and benefits to different social groups, which in turn has consequences for political party competition. Part of the populist appeal can be interpreted as a representation of resentment against the educated middle classes, who either help design the regulatory framework, or else are engaged in citizen expertise, activities from which they benefit, or which chimes with their ideology. The much-celebrated 'co-production of science and society' has its blind spot in the excluded who are not involved in the co-producing.

This structural change is accompanied by a proliferation of technological expert systems which have taken over tasks from technical specialists but also extended the range of possible new applications. While the capabilities of such systems have been hyped, they are here to stay, too. They have some causal agency but lack moral responsibility and legal liability. But political decisions could enact legislation that regulate their use and scope.

Democratic societies need to find ways to deal with these challenges. They need to be prepared to open up technical black boxes that are allegedly inevitable and innovative, or have been unproblematic in the past. Political power is about deciding which issues are allowed to become politicized, and which are scientized. Enabling citizen expertise, and opening up technical debates to interventions from below are vital in this regard.

However, democratic societies also need to establish fair procedures that recognize the uneven access to knowledge and its social representation. Social groups that cannot, or do not want to articulate their grievances through forms of citizen expertise may nevertheless do so at the polling stations. Patronizing them from a position of knowing better will only deepen the divide.

Finally, the regulation of global information companies, and AI, is something that cannot be ignored. The march of AI is not inevitable, and there are other options than embracing their promise or rejecting them as dystopian. AI is not a fateful event or process, but the result of corporate strategies, regulatory frameworks, public attitudes, and consumer needs. The question is not so much if we should be frightened or relaxed. Rather, we should ask: what kind of regulation do we deem necessary? What scope should it have, who should be involved in deciding? Which kind of society do we want to live in? At this point it becomes evident, once more, that expertise is a complex matter. It is the product of social and political forces, and is in turn influencing society. The aim should be to base any political intervention on a profound understanding of expertise in a rapidly changing world.

Notes

1 Media coverage tends to follow announcements from private companies developing AI, as Brennen et al. (2018) show. They also point out that the connections between AI coverage and technology reporting practices should be interpreted in the light of 'structural changes seen across health, science, and technology journalism over the past two decades.' As a result of financial pressures, many news outlets are no longer in a position to carry out in-depth analysis or investigative work. More and more they have come to 'rely heavily on press releases for day-to-day science and technology news stories' (Brennen et al. 2018:8). This means the way we perceive AI is likely to be framed by the companies who are most active in its development.

2 Developments in the past years seem to cast doubt on the general validity of this statement. The financial crisis exemplifies the current state of affairs in which the technocracy did not 'dope out' the public before it became fully aware. Financial experts were as surprised as the allegedly uninformed public. There are other prominent examples of such moments of surprise, such as the fall of the Communist bloc, the discovery of the ozone hole, Brexit, or the election of Donald Trump. Nevertheless, some 'new populists' allege

intention on the part of experts in 'headquarters' to develop their hidden plans, 'doping out' the public.

3 How enthusiastically Bell embraced the power of technocracy can be seen from this comment: 'Daniel Bell, for instance, endorsed the idea of leadership by a "scientific elite." … People like Bell, in other words, would help society manage the transition to a post-industrial society; having the ear of the sovereign, they would offer a critical perspective on the impact and use of new technology. Whatever critical perspective they might possess, intellectuals like Bell and McLuhan in the 1960s and 1970s were frequently on the payroll of government or business – and often both, whether as consultants, contractors or speakers' (Cummings 2016:17).

4 It is beyond the scope of this chapter to examine Habermas' theory in more detail. In this context, it should be noted that he replaced the Old Frankfurt School criticism of Western rationalism with a view inspired by American pragmatism. Initially, this was done through the distinction between system and lifeworld, where the system is the necessary (!) realm of instrumental rationality and efficiency and the lifeworld is the realm where meaningful interaction between citizens takes place (aka 'communicative action').

5 Anyone who has experienced 'home office working' ca. 2020 will concur that Luhmann's point seems obvious.

References

Abbott, Andrew. 1988. *The System of Professions: An Essay on the Division of Expert Labor*. Chicago: University of Chicago Press.

Adam, David. 2022. "The Pandemic's True Death Toll: Millions More than Official Counts." *Nature* 601(7893): 312–15.

Adler, Emanuel, and Peter M. Haas. 1992. "Conclusion: Epistemic Communities, World Order, and the Creation of a Reflective Research Program." *International Organization* 46(1): 367–90.

Adolf, Marian, and Nico Stehr. 2017. *Knowledge*. Second ed. London: Routledge.

Agrawala, Shardul. 1998. "Context and Early Origins of the Intergovernmental Panel on Climate Change." *Climatic Change* 39: 605–20.

Ahmed, Ayesha, Prabadevi Boopathy, and S. Sudhagara Rajan. 2022. "Artificial Intelligence for the Novel Corona Virus (COVID-19) Pandemic: Opportunities, Challenges, and Future Directions." *International Journal of E-Health and Medical Communications* 13(2): 1–21.

Akerlof, George A. 1970. "The Market for 'Lemons': Quality Uncertainty and the Market Mechanism." *The Quarterly Journal of Economics* 84(3): 488.

Akrich, Madeleine, and Vololona Rabeharisoa. 2012. "L'expertise profane dans les associations de patients, un outil de démocratie sanitaire." *Santé Publique* 24(1): 69–74.

Alwan, Nisreen A. et al. 2020. "Scientific Consensus on the COVID-19 Pandemic: We Need to Act Now." *The Lancet*, 396: e71–e72.

Anderson, Chris. 2008. "The End of Theory: The Data Deluge Makes the Scientific Method Obsolete." *WIRED*. https://www.wired.com/2008/06/pb-theory/ (November 25, 2021).

Andresen, Steinar, and Tora Skodvin. 2011. "The Climate Regime: Achievements and Challenges." In *The World Ocean in Globalisation: Climate Change, Sustainable Fisheries, Biodiversity, Shipping, Regional Issues*, eds. Davor Vidas and Peter Johan Schei. Fridtjof Nansen Institute, Lysaker, Norway: Brill, 165–86.

Angwin, Julia, Jeff Larson, Surja Mattu, and Lauren Kirchner. 2016. "Machine Bias: There's Software Used across the Country to Predict Future Criminals. And It's Biased against Blacks." *ProPublica*. https://www.propublica.org/article/machine-bias-risk-assessments-in-criminal-sentencing (November 25, 2021).

Arksey, Hilary. 1994. "Expert and Lay Participation in the Construction of Medical Knowledge." *Sociology of Health & Illness* 16(4): 448–68.

Autor, David H. 2015. "Why Are There Still so Many Jobs? The History and Future of Workplace Automation." *Journal of Economic Perspectives* 29(3): 3–30.

Baldwin, Mark P., and Timothy M. Lenton. 2020. "Solving the Climate Crisis: Lessons from Ozone Depletion and COVID-19." *Global Sustainability*3: 1–6.

Banks, Melanie. 2021. "The Paris Agreement: Truth, Obligations and 'Noble Lies.'" *Ethics & the Environment*: 1–20. https://muse-jhu-edu.nottingham.idm.oclc.org/article/837716 (January 10, 2022).

Barker, Kristin K. 2008. "Electronic Support Groups, Patient-Consumers, and Medicalization: The Case of Contested Illness." *Journal of Health and Social Behavior* 49(1): 20–36.

Baumol, William, and William G. Bowen. 1965. "On the Performing Arts: The Anatomy of Their Economic Problems." *The American Economic Review* 55(1/2): 495–502.

Bear, Laura. 2020. "Covid Rules Need to Put People First." *The Guardian*. 27 October

Beasley, Ryan A. 2012. "Medical Robots: Current Systems and Research Directions." *Journal of Robotics*: 1–14.

Beck, Silke. 2012. "Between Tribalism and Trust: The IPCC Under the 'Public Microscope.'" *Nature and Culture* 7(2): 151–73.

Beck, Silke, and Martin Mahony. 2017. "The IPCC and the Politics of Anticipation." *Nature Climate Change* 7(5): 311–13.

———. 2018. "The IPCC and the New Map of Science and Politics." *Wiley Interdisciplinary Reviews: Climate Change* 9(6): e547.

Bell, Daniel. 1973. *The Coming of Post-Industrial Society*. New York: Basic Books.

Bell, Stephen, and Andrew Hindmoor. 2014. "The Structural Power of Business and the Power of Ideas: The Strange Case of the Australian Mining Tax." *New Political Economy* 19(3): 470–86.

Berg, Monika, and Rolf Lidskog. 2018. "Pathways to Deliberative Capacity: The Role of the IPCC." *Climatic Change*. 148:11–24.

Berger, Peter, and Thomas Luckmann. 1966. *The Social Construction of Reality. A Treatise in the Sociology of Knowledge*. London: Penguin Books.

Bermeo, Nancy. 2016. "On Democratic Backsliding." *Journal of Democracy* 27(1): 5–19.

Berrang-Ford, Lea et al. 2021. "A Systematic Global Stocktake of Evidence on Human Adaptation to Climate Change." *Nature Climate Change* 11(11): 989–1000.

Bhorat, Zyyad. 2017. "Do We Still Need Human Judges in the Age of Artificial Intelligence?" *openDemocracy*. https://www.opendemocracy.net/en/transformation/do-we-still-need-human-judges-in-age-of-artificial-intelligence/ (November 30, 2019).

Blicharska, Malgorzata et al. 2017. "Steps to Overcome the North-South Divide in Research Relevant to Climate Change Policy and Practice." *Nature Climate Change* 7(1): 21–7.

Bodansky, Daniel. 1993. "The United Nations Framework Convention on Climate Change: A Commentary." *Yale Journal of International Law* 18(3): 451–558.

Boehmer-Christiansen, Sonja. 1994a. "Global Climate Protection Policy: The Limits of Scientific Advice, Part 1." *Global Environmental Change* 4(2): 140–59.

———. 1994b. "Global Climate Protection Policy: The Limits of Scientific Advice, Part 2." *Global Environmental Change* 4(3): 185–200.

Böhme, Gernot, and Nico. Stehr. 1986. *The Knowledge Society : The Growing Impact of Scientific Knowledge on Social Relations*. Dordrecht, the Netherlands: Reidel Pub. Co.

Borie, Maud, Mark Pelling, Gina Ziervogel, and Keith Hyams. 2019. "Mapping Narratives of Urban Resilience in the Global South." *Global Environmental Change* 54: 203–13.

Borie, Maud, Martin Mahony, Noam Obermeister, and Mike Hulme. 2021. "Knowing like a Global Expert Organization: Comparative Insights from the IPCC and IPBES." *Global Environmental Change* 68(May 2020): 102261.

Bowker, Geoffrey C., and Susan Leigh Star. 1999. *Sorting Things out: Classification and Its Consequences*. Cambridge, MA: MIT Press.

Bowman, Shayne, and Chris Willis. 2003. 'We Media: How Audiences Are Shaping the Future of News and Information.' The Media Centre Thinking Paper. https://www.hypergene.net/wemedia/download/we_media.pdf

Brake, David R. 2014. "Are We All Online Content Creators Now? Web 2.0 and Digital Divides." *Journal of Computer-Mediated Communication* 19(3): 591–609.

Branswell, Helen. 2022. "WHO Official Weighs in on Covid, Vaccines, and Mistakes That Were Made." *Statnews*. https://www.statnews.com/2022/01/03/a-who-official-weighs-in-on-covid-vaccines-and-mistakes-that-were-made/ (January 4, 2022).

Brants, Kees, and Jo Bardoel. 2008. "Death Duties: Kelly, Fortuyn and Their Challenge to Media Governance." *European Journal of Communication* 23(4): 471–89.

Brennan, Jason. 2005. *Against Democracy*. Princeton, NJ: Princeton University Press.

Brennen, J. Scott, Philip N. Howard, and Rasmus Kleis Nielsen. 2018. "An Industry-Led Debate: How UK Media Cover Artificial Intelligence." Working Paper, Reuters Institute for the Study of Journalism as part of the Oxford Martin Programme on Misinformation, Science and Media

Brilliant, Larry et al. 2021. "The Forever Virus. A Strategy for the Long Fight Against COVID-19." *Foreign Affairs*. July-August 2021.

Brooks, Thom. 2004. "The Right to Trial by Jury." *Journal of Applied Philosophy* 21(2): 197–212.

Brown, Gordon, and Daniel Susskind. 2020. "International Cooperation during the COVID-19 Pandemic." *Oxford Review of Economic Policy* 36: S64–76.

Brown, Jennings. 2017. "Why Everyone Is Hating on IBM Watson—Including the People Who Helped Make It." https://gizmodo.com/why-everyone-is-hating-on-watson-including-the-people-w-1797510888 (November 25, 2021).

Brown, Mark B. 2009. *Science in Democracy: Expertise, Institutions, and Representation*. Cambridge, MA: The MIT Press.

Buck, Holly, Oliver Geden, Masahiro Sugiyama, and Olaf Corry. 2020. "Pandemic Politics—Lessons for Solar Geoengineering." *Communications Earth & Environment* 1(1): 16.

Burgman, Mark A. 2016. *Trusting Judgements*. Cambridge: Cambridge University Press.

Burrell, Jenna. 2016. "How the Machine 'Thinks': Understanding Opacity in Machine Learning Algorithms." *Big Data and Society* 3(1): 1–12.

Busch, Andreas. 2009. "Politikwissenschaft und Politikberatung: Reflektionen Anlässlich Der Aktuellen Krise." *Zeitschrift für Politikberatung* 2(3): 467–84.

Butos, William N., and Thomas J. McQuade. 2015. "Causes and Consequences of the Climate Science Boom." *Independent Review* 20(2): 165–96.

Cairney, Paul. 2021. "The UK Government's COVID-19 Policy: What Does 'Guided by the Science' Mean in Practice?" *Frontiers in Political Science* 3(March): 1–14.

Callard, Felicity, and Elisa Perego. 2021. "How and Why Patients Made Long Covid." *Social Science & Medicine* 268: 113426.

Callon, M. 1999. "The Role of Lay People in the Production and Dissemination of Scientific Knowledge." *Science Technology & Society* 4(1): 81–94.

Cambrosio, Alberto, Camille Limoges, and Eric Hoffman. 1992. "Expertise as a Network: A Case Study of the Controversies over the Environmental Release of Genetically Engineered Organisms." In *The Culture and Power of Knowledge: Inquiries into Contemporary Societies*, eds. Nico Stehr and Richard V,. Ericson. 341–62.

Campbell, John L., and Ove K. Pedersen. 2010. "Knowledge Regimes and Comparative Political Economy." In *Ideas and Politics in Social Science Research*, eds. Daniel Béland and Robert Henry Cox. New York, NY: Oxford University Press, 167–190.

Campbell, John L., and Ove K. Pederson. 2014. *Knowledge Regimes and the National Origins of Policy Ideas. Knowledge Regimes in the United States, France, Germany, and Denmark*. Princeton, NJ: Princeton University Press.

Campbell, John L., and Ove K. Pedersen. 2015. "Policy Ideas, Knowledge Regimes and Comparative Political Economy." *Socio-Economic Review* 13(4): 679–701.

Carlson, Matt. 2015. "The Robotic Reporter: Automated Journalism and the Redefinition of Labor, Compositional Forms, and Journalistic Authority." *Digital Journalism* 3(3): 416–31.

Celermajer, Danielle, and Dalia Nassar. 2020. "COVID and the Era of Emergencies: What Type of Freedom Is at Stake?" *Democratic Theory* 7(2): 12–24.

Clerwall, Christer. 2014. "Enter the Robot Journalist." *Journalism Practice* 8(5): 519–31.

Clough, Paul. 2020. "Why You Should Never Use Microsoft Excel to Count Coronavirus Cases." *The Conversation*. https://theconversation.com/why-you-should-never-use-microsoft-excel-to-count-coronavirus-cases-147681 (October 13, 2020).

Cobbe, Jennifer. 2019. "Administrative Law and the Machines of Government: Judicial Review of Automated Public-Sector Decision-Making." *Legal Studies*. 39(4): 636–655.

Coglianese, Cary, and David Lehr. 2017. "Regulating by Robot: Administrative Decision Making in the Machine-Learning Era." *Georgetown Law Journal* 105(5): 1147–1223.

Collingridge, David, and Colin Reeve. 1986. *Science Speaks to Power*. New York: St Martin's Press.

Collins, Harry. 1985. Changing Order: Replication and Induction in Scientific Practice. London: Sage.

———. 2014. Are We All Scientific Experts Now? Cambridge: Polity Press.

———. 2007. Rethinking Expertise. Chicago, IL: University of Chicago Press.

Collins, Harry, and Rob Evans. 2002. "The Third Wave of Science Studies: Studies of Expertise and Experience." *Social Studies of Science* 32(2): 235–96.

———. 2007. *Rethinking Expertise*. Chicago, IL: University of Chicago Press.

Collins, Harry, Robert Evans, and Martin Weinel. 2017. "STS as Science or Politics?" *Social Studies of Science*: 47(4): 580–586.

Connolly, William E. 2002. *Identity, Difference: Democratic Negotiations of Political Paradox*. Minneapolis, MN: Minnesota University Press.

Cook, John et al. 2013. "Quantifying the Consensus on Anthropogenic Global Warming in the Scientific Literature." *Environmental Research Letters* 8(2): 024024.

Cooper, Ryan. 2021. "Noble Lies Are a Public Health Hazard." *The Week*: 17 December. https://theweek.com/coronavirus/1008155/noble-lies-are-a-public-health-hazard (January 11, 2022).

Corbera, Esteve, Laura Calvet-Mir, Hannah Hughes, and Matthew Paterson. 2016. "Patterns of Authorship in the IPCC Working Group III Report." *Nature Climate Change* 6(1): 94–99.

Corfee-Morlot, Jan, Mark Maslin, and Jacquelin Burgess. 2007. "Global Warming in the Public Sphere." *Philosophical Transactions of the Royal Society* 365(July): 2741–76.

Cummings, Alex Sayf. 2016. "Of Sorcerers and Thought Leaders: Marketing the Information Revolution in the 1960s." *Sixties* 9(1): 1–25.

Cummings, Dominic. 2017. "How the Brexit Referendum Was Won." *The Spectator*. https://blogs.spectator.co.uk/2017/01/dominic-cummings-brexit-referendum-won/ (January 23, 2020).

Cushion, Stephen, Justin Lewis, and Gordon Neil Ramsay. 2012. "The Impact of Interventionist Regulation in Reshaping News Agendas: A Comparative Analysis of Public and Commercially Funded Television Journalism." *Journalism* 13(7): 831–49.

Daston, Lorraine. 2020. "Ground-Zero Empiricism." *Critical Enquiry* 47(S2): S55–7.

Daum, Mathias. 2020. "Schön Locker und Sorglos." *Die Zeit*.

Davidson, James Dale, and William Rees-Mogg. 1997. *The Sovereign Individual: How to Survive and Thrive during the Collapse of the Welfare State.* New York: Simon & Schuster.

De Nicola, Giacomo, Göran Kauermann, and Michael Höhle. 2021. "On Assessing Excess Mortality in Germany during the COVID-19 Pandemic." *AStA Wirtschafts- und Sozialstatistisches Archiv 2022*: 29(3):3–30.

De Tocqueville, Alexis. 1835. *Democracy in America.* Pennsylvania State University (2002).

De Vries, Gerard. 2007. "What Is Political in Sub-Politics? How Aristotle Might Help STS." *Social Studies of Science* 37(5): 781–809.

Detenber, Benjamin H. et al. 2014. "Rights Versus Morality: Online Debate About Decriminalization of Gay Sex in Singapore." *Journal of Homosexuality* 61(9): 1313–33.

Dowd, Jennifer Beam et al. 2020. "Demographic Science Aids in Understanding the Spread and Fatality Rates of COVID-19." *Proceedings of the National Academy of Sciences of the United States of America* 117(18): 9696–8.

Dressel, Julia, and Hany Farid. 2018. "The Accuracy, Fairness, and Limits of Predicting Recidivism." *Science Advances* 4(1): 1–6.

Dryzek, John S. 2002. *Deliberative Democracy and Beyond.* Oxford: Oxford University Press.

Edenhofer, Ottmar, and Jan Minx. 2014. "Mapmakers and Navigators, Facts and Values." *Science* 345(6192): 37–8.

Edwards, Paul N. 2010. *A Vast Machine.* Cambridge, MA: The MIT Press.

Elden, Stuart. 2007. "Governmentality, Calculation, Territory." *Environment and Planning D: Society and Space* 25(3): 562–80.

Engler, Sarah et al. 2021. "Democracy in Times of the Pandemic: Explaining the Variation of COVID-19 Policies across European Democracies." *West European Politics* 44(5–6): 1077–102.

Entman, R.M. 1993. "Framing: Towards Clarification of a Fractured Paradigm." *Journal of Communication* 43(4): 51–58.

Epstein, Steven. 1995. "The Construction of Lay Expertise: AIDS Activism and the Forging of Credibility in the Reform of Clinical Trials." *Science, Technology & Human Values* 20(4): 408–37.

———. 1996. *Impure Science : AIDS, Activism, and the Politics of Knowledge.* Berkeley, CA: University of California Press.

Erikainen, Sonja, and Ellen Stewart. 2020. "Credibility Contests: Media Debates on Do-It-Yourself Coronavirus Responses and the Role of Citizens in Health Crises." *Frontiers in Sociology* 5: 91.

Evans, Simon et al. 2021. "In-Depth Q&A: The IPCC's Sixth Assessment Report on Climate Science—Carbon Brief." *Carbonbrief.* https://www.carbonbrief.org/in-depth-qa-the-ipccs-sixth-assessment-report-on-climate-science (January 20, 2022).

Eyal, Gil. 2013. "For a Sociology of Expertise: The Social Origins of the Autism Epidemic." *American Journal of Sociology* 118(4): 863–907.

———. 2019. *The Crisis of Expertise.* Cambridge: Polity Press.

———. 2020. "Beware the Trolley Zealots." *Sociologica* 14(1): 21–30.

Falzon, Danielle. 2021. "Expertise and Exclusivity in Adaptation Decision-Making." *Current Opinion in Environmental Sustainability* 51: 95–100.

Farrell, David M., Jane Suiter, and Clodagh Harris. 2019. "'Systematizing' Constitutional Deliberation: The 2016–18 Citizens' Assembly in Ireland." *Irish Political Studies* 34(1): 113–23.

Ferrucci, David et al. 2010. "Building Watson: An Overview of the DeepQA Project." *AI Magazine* 31(3): 59.

Fischer, Frank. 2003. *Reframing Public Policy: Discursive Politics and Deliberative Practices.* Oxford: Oxford University Press.

Fiske, Shirly J. et al. 2014. "Changing the Atmosphere. Anthropology and Climate Change. *Final Report of the AAA Global Climate Change Task Force*, p. 137. December 2014. Arlington, VA: American Anthropological Association."

Foerster, Heinz von. 1981. *Observing Systems*. Seaside, CA: Intersystems Publications.

Forrest, Katherine B. 2021. *When Machines Can Be Judge, Jury, And Executioner: Justice In The Age Of Artificial Intelligence*. Singapore: World Scientific Publishing. 1–134.

Foucault, Michel. 1980. *Power/Knowledge: Selected Interviews and Other Writings, 1972–1977*. New York: Vintage.

Franzosi, Roberto. 1998. "Narrative Analysis—Or Why (and How) Sociologists Should Be Interested in Narrative." *Annual Review of Sociology* 24: 517–54.

Freedland, Jonathan. 2016. "Post-Truth Politicians Such as Donald Trump and Boris Johnson Are No Joke." *The Guardian*, 13 May.

Frey, Carl Benedikt, and Michael A. Osborne. 2013. *The Future of Employment*. Oxford: Oxford Martin School.

Fromm, Erich. 1992. *The Art of Being*. New York: Open Road Integrated Media, Inc.

Fukuyama, Francis. 1995. *Trust: The Social Virtues and the Creation of Prosperity*. New York: The Free Press.

Fuller, Steve. 2017. "Is STS All Talk and No Walk?" *EASST Review* 36(1).

———. 2018. *Post-Truth: Knowledge As A Power Game*. New York: Anthem Press.

———. 2021. "Democracy Naturalized: In Search of the Individual in the Post-Truth Condition." *Analyse & Kritik* 43(2): 351–66.

Funtowicz, Silvio O., and Jerome Ravetz. 1990. *Uncertainty and Quality in Science for Policy*. Dordrecht, the Netherlands: Kluwer Academic Publishers.

Funtowicz, Silvio O., and Jerome R. Ravetz. 1993. "Science for the Post-Normal Age." *Futures* 25(7): 739–55.

Fürst, Silke, Philomen Schönhagen, and Stefan Bosshart. 2015. "Mass Communication Is More than a One-Way Street: On the Persistent Function and Relevance of Journalism." *Javnost* 22(4): 328–44.

Ganesh, Janan. 2021. "Vaccination and the Bonfire of Asian Clichés." *Financial Times*. 30 July.

Gardiner, Michael E. 2012. *Weak Messianism: Essays in Everyday Utopianism*. Bern: Peter Lang.

Gava, John. 2009. "A Study in Judging: Sir Owen Dixon and McDonald v Dennys Lascelles." *Australian Bar Review* 32(1): 77–95.

Geden, Oliver. 2018. "Politically Informed Advice for Climate Action." *Nature Geoscience* 11(6): 380–83.

Gelbspan, Ross. 1997. *The Heat Is on: The Climate Crisis, the Cover-up, the Prescription*. New York: Perseus Books.

Gerhards, Jürgen. 2001. "Der Aufstand Des Publikums: Eine Systemtheoretische Interpretation Des Kulturwandels in Deutschland Zwischen 1960 Und 1989." *Zeitschrift für Soziologie* 30(3): 163–84.

Gieryn, Thomas F. 1995. "Boundaries of Science." In *Handbook of Science and Technology Studies*, eds. Sheila Jasanoff, Gerald E. Markle, James C. Peterson, and Trevor Pinch. London: Sage, 392–443.

Gigerenzer, Gerd et al. 2005. "'A 30% Chance of Rain Tomorrow': How Does the Public Understand Probabilistic Weather Forecasts?" *Risk Analysis* 25(3): 623–9.

———. 2022. *How to Stay Smart in a Smart World*. London: Allen Lane.

Goldenberg, Maya. 2021. *Vaccine Hesitancy: Public Trust, Expertise, and the War on Science*. Pittsburgh: University of Pittsburgh Press.

Gonçalves, Maria Eduarda. 1995. "Scientific Expertise and European Community Regulatory Processes." *Science and Public Policy* 22(3): 183–7.

Goode, Luke. 2009. "Social News, Citizen Journalism and Democracy." *New Media and Society* 11(8): 1287–1305.

Graefe, Andreas, Mario Haim, Bastian Haarmann, and Hans Bernd Brosius. 2018. "Readers' Perception of Computer-Generated News: Credibility, Expertise, and Readability." *Journalism* 19(5): 595–610.

Granovetter, Mark. 1978. "Threshold Models of Collective Behavior." *American Journal of Sociology* 83(6): 1420–43.

Grove, William M., and Paul E. Meehl. 1996. "Comparative Efficiency of Informal (Subjective, Impressionistic) and Formal (Mechanical, Algorithmic) Prediction Procedures: The Clinical-Statistical Controversy." *Psychology, Public Policy, and Law* 2(2): 293–323.

Grundmann, Reiner. 1999. "The European Public Sphere and the Deficit of Democracy." eds. Dennis Smith and Sue Wright. *Sociology Review Monograph*: 125–46.

———. 2007. "Climate Change and Knowledge Politics." *Environmental Politics* 16(3): 414–32.

———. 2008. "What May the Sheep Safely Know?" In *Knowledge & Democracy: A 21st-Century Perspective*, ed. Nico Stehr. New Brunswick, NJ: Transaction Publishers, 85–103.

———. 2012. "'Climategate' and the Scientific Ethos." *Science, Technology & Human Values* 38(1): 67–93.

———. 2013. "Debunking Sceptical Propaganda." *BioSocieties* 8(3): 370–74.

———. 2015. "Climate Skepticism." In *Research Handbook on Climate Governance*, eds. Eva Lovbrand and Karen Backstrand. Cheltenham: Edward Elgar, 175–88.

———. 2016. "Climate Change as a Wicked Social Problem." *Nature Geosciences* 9: 562–63.

———. 2017. "The Problem of Expertise in Knowledge Societies." *Minerva* 55(1): 25–48.

———. 2018. "The Rightful Place of Expertise." *Social Epistemology* 32(6): 372–86.

———. 2021. "COVID and Climate: Similarities and Differences." *WIREs Climate Change* 12(6).

Grundmann, Reiner, and Christos Mantziaris. 1991. "Fundamentalist Intolerance or Civil Disobedience? Strange Loops in Liberal Theory." *Political Theory* 19(4): 572–605.

Grundmann, Reiner, and Simone Rödder. 2019. "Sociological Perspectives on Earth System Modelling." *Journal of Advances in Modeling Earth Systems* 11(12): 3878–92.

Grundmann, Reiner, and Nico Stehr, eds. 2005. *Knowledge (Critical Concepts in Philosophy)*. Five Volumes. London: Routledge.

———. 2012. *The Power of Scientific Knowledge: From Research to Public Policy*. Cambridge: Cambridge University Press.

Gu, Youyang et al. 2021. "21 Experts on the Future of Expertise." *Future a16z*. https://future.a16z.com/question/future-of-expertise/ (November 10, 2021).

Guillemot, Hélène. 2017. "The Necessary and Inaccessible 1.5°C Objective: A Turning Point in the Relations between Climate Science and Politics?" In *Globalising the Climate: COP21 and the Climatisation of Global Debates*, eds. Stefan C. Aykut, Jean Foyer, and Edouard Morena. London: Routledge, 39–56.

Gunderson, Ryan. 2018. "Global Environmental Governance Should Be Participatory: Five Problems of Scale." *International Sociology* 33(6): 715–37.

Guston, David H. 2001. "Boundary Organizations in Environmental Policy and Science: An Introduction." *Science, Technology, & Human Values* 26(4): 399–408.

Gutiérrez, María, James Van Alstine, and Yulia Yamineva. 2011. 12 *Earth Negotiations Bulletin Summary of the 34th Session of the Intergovernmental Panel on Climate Change*: 18–19, November 2011.

Haas, Peter M. 1992. "Banning Chlorofluorocarbons: Epistemic Community Efforts to Protect Stratospheric Ozone." *International Organization* 46(1): 187–224.

Haas, Peter M. 1993. Stratospheric ozone: Regime formation in stages. In *Polar politics: Creating international environmental regimes*, edited by O. Young and G. Osherenko. Ithaca, NY: Cornell University Press, 152–85.

Haas, Peter M. 2004. "When Does Power Listen to Truth? A Constructivist Approach to the Policy Process." *Journal of European Public Policy* 11(4): 569–92.

Habermas, Jürgen. 1971a. "Technology and Science as Ideology." In *Toward a Rational Society: Student Protest, Science, and Politics*, Boston: Beacon Press, 81–122.

———. 1971b. "The Scientization of Politics and Public Opinion." In *Toward a Rational Society: Student Protest, Science, and Politics*, Boston, MA: Beacon Press, 62–80.

Habersaat, Katrine Bach et al. 2020. "Ten Considerations for Effectively Managing the COVID-19 Transition." *Nature Human Behaviour* 4(7): 677–87.

Hackett, Paul, and Paul Hunter. 2015. *Who Governs Britain ? A Profile of MPs in the New Parliament*. London: The Smith Institute.

Hackmann, Heide, Susanne C. Moser, and Asuncion Lera St. Clair. 2014. "The Social Heart of Global Environmental Change." *Nature Climate Change* 4(8): 653–5.

Hajer, Maarten A. 1995. *The Politics of Environmental Discourse: Ecological Modernization and the Policy Process*. Oxford: Clarendon Press.

Hall, P.A. 1993. "Policy Paradigms, Social Learning, and the State: The Case of Economic Policymaking in Britain." *Comparative politics* 25(3): 275–96.

Hanna, Meena. 2019. "Robo-Judge: Common Law Theory and the Artificially Intelligent Judiciary." *Journal of Judicial Administration* 29(1): 22–42.

Hans, Valerie P., and Neil Vidmar. 1986. "From Trial by Ordeal to Trial by Jury." Eds. Hans, Valerie P. and Vidmar, Neil. *Judging the Jury*, Boston, MA.: Springer, 21–30.

Hart, David M., and David G. Victor. 1993. "Scientific Elites and the Making of US Policy for Climate Change Research, 1957–74." *Social Studies of Science* 23(4): 643–80.

Harvard Law Review. 2017. "State v. Loomis." *Harvard Law Review*. https://harvardlawreview.org/2017/03/state-v-loomis/ (December 2, 2021).

Hasen, Richard L. 2012. "Lobbying, Rent-Seeking, and the Constitution." *Stanford Law Review* 64(1): 191–253.

Haunschild, Robin, Lutz Bornmann, and Werner Marx. 2016. "Climate Change Research in View of Bibliometrics." *PloS one* 11(7): e0160393.

Heymann, Matthias. 2018. "The Climate Change Dilemma: Big Science, the Globalizing of Climate and the Loss of the Human Scale." *Regional Environmental Change*: 19:1549–1560.

Ho, C. W. L., D. Soon, K. Caals, and J. Kapur. 2019. "Governance of Automated Image Analysis and Artificial Intelligence Analytics in Healthcare." *Clinical Radiology* 74(5): 329–37.

Hobson-West, Pru. 2003. "Understanding Vaccination Resistance: Moving beyond Risk." *Health, Risk & Society* 5(3): 273–83.

Hoffmeister, Thaddeus, and Ann Charles Watts. 2018. "Social Media, the Internet, and Trial by Jury." *Annual Review of Law and Social Science* 14(1): 259–70.

Hollin, G. J. S., and W. Pearce. 2015. "Tension between Scientific Certainty and Meaning Complicates Communication of IPCC Reports." *Nature Climate Change* 5(8): 753–6.

Holm, Poul, and Verena Winiwarter. 2017. "Climate Change Studies and the Human Sciences." *Global and Planetary Change* 156(July 2016): 115–22.

Hoppe, Imke, and Simone Rödder. 2019. "Speaking with One Voice for Climate Science—Climate Researchers' Opinion on the Consensus Policy of the IPCC." *Journal of Science Communication* 18(03): A04.

Hoppe, R., and A. Wesselink. 2014. "Comparing the Role of Boundary Organizations in the Governance of Climate Change in Three EU Member States." *Environmental Science and Policy* 44: 73–85.

Hoppe, Rob, Anna Wesselink, and Rose Cairns. 2013. "Lost in the Problem: The Role of Boundary Organisations in the Governance of Climate Change." *Wiley Interdisciplinary Reviews: Climate Change* 4(4): 283–300.

Hoppit, Julian. 1996. "Political Arithmetic in Eighteenth-Century England." *Economic History Review* 49(3): 516–40.

Horton, Richard. 2020. "Global Health: Time for Radical Change?" *The Lancet* 396:1129.

Høstaker, Roar. 2005. "Latour—Semiotics and Science Studies." *Science & Technology Studies* 18(2): 5–25.

Howarth, Candice, and James Painter. 2016. "Exploring the Science–Policy Interface on Climate Change: The Role of the IPCC in Informing Local Decision-Making in the UK." *Palgrave Communications* 2(1): 1–12.

Hughes, Hannah. 2015. "Bourdieu and the IPCC's Symbolic Power." *Global Environmental Politics* 15(4): 85–104.

Hughes, Hannah Rachel, and Matthew Paterson. 2017. "Narrowing the Climate Field: The Symbolic Power of Authors in the IPCC's Assessment of Mitigation." *Review of Policy Research* 34(6): 744–66.

Hulme, Mike. 2010. "Problems with Making and Governing Global Kinds of Knowledge." *Global Environmental Change* 20(4): 558–64.

Hulme, Mike, Rolf Lidskog, James M. White, and Adam Standring. 2020. "Social Scientific Knowledge in Times of Crisis: What Climate Change Can Learn from Coronavirus (and Vice Versa)." *Wiley Interdisciplinary Reviews: Climate Change* 11(4): 1–5.

Hulme, Mike, and Martin Mahony. 2010. "Climate Change : What Do We Know about the IPCC?" *Progress in Physical Geography*: 34(5): 705–718.

Huntingford, Chris et al. 2019. "Machine Learning and Artificial Intelligence to Aid Climate Change Research and Preparedness." *Environmental Research Letters* 14(12): 124007.

Illich, Ivan. 1973. *Tools for Conviviality*. London: Calder and Boyars.

Iosifidis, Petros, and Leighton Andrews. 2019. "Regulating the Internet Intermediaries in a Post-Truth World: Beyond Media Policy?" *International Communication Gazette*: 82(3): 211–230.

IPCC. 2003. *Report of the Twentieth Session of the Intergovernmental Panel on Climate Change (IPCC) Paris*, 19–21 February 2003.

———. 2019. *Special Report: The Ocean and Cryosphere in a Changing Climate*. https://www.ipcc.ch/srocc/

IPCC AR4 Synthesis Report. 2007. Intergovernmental Panel on Climate Change *Intergovernmental Panel on Climate Change. Fourth Assessment Report*. Cambridge: Cambridge University Press.

IPCC, 2015. Expert Meeting Report of the Intergovernmental Panel on Climate Change on Potential Studies of the IPCC Process [Sokona, Y.; Calvo, E.; Christ, R.; Johnson, C.; Oreskes, N.; Witi, J. and Wratt, D. (eds.)]. IPCC Secretariat c/o WMO, Geneva, Switzerland, pp. 41

Irwin, Alan. 1995. *Citizen Science: A Study of People, Expertise and Sustainable Development*. London: Routledge.

———. 2017. "Agreeing to Differ? A Response to Van Bouwel and Van Oudheusden." *Social epistemology Review and reply Collective* 10(10): 11–14.

Jacobsen, John Kurt. 1995. "Much Ado About Ideas: The Cognitive Factor in Economic Policy." *World Politics* 47(2): 283–310.

Janse, Maartje. 2019. "What Value Should We Attach to All These Petitions?: Petition Campaigns and the Problem of Legitimacy in the Nineteenth-Century Netherlands." *Social Science History* 43(3): 509–30.

Jasanoff, Sheila. 1990. *The Fifth Branch: Science Advisers as Policymakers*. Cambridge, MA: Harvard University Press.

———. 2004. *States of Knowledge: The Co-Production of Science and the Social Order*. London: Routledge.

———. 2005. *Designs on Nature: Science and Democracy in Europe and the United States*. Princeton, NJ: Princeton University Press.

———. 2011a. "Cosmopolitan Knowledge: Climate Science and Global Civic Epistemology." In *The Oxford Handbook of Climate Change and Society*, eds. Dryzek, John S., Norgaard, Richard B., and David Schlosberg. Oxford: Oxford University Press, 129–43.

———. 2011b. "Quality Control and Peer Review in Advisory Science." In *The Politics of Scientific Advice. Institutional Design for Quality Assurance*, eds. Peter Weingart and Justus Lentsch. Cambridge: Cambridge University Press, 19–35.

Jasanoff, Sheila, and Hilton R. Simmet. 2017. "No Funeral Bells: Public Reason in a 'Post-Truth' Age." *Social Studies of Science* 47(5): 751–70.

Johnston, Lisette. 2016. "Social News = Journalism Evolution?: How the Integration of UGC into Newswork Helps and Hinders the Role of the Journalist." *Digital Journalism* 4(7): 899–909.

Jones, Michael D. 2014. "Cultural Characters and Climate Change: How Heroes Shape Our Perception of Climate Science." *Social Science Quarterly* 95(1): 1–39.

Jones, N. 2017. "How Machine Learning Could Help to Improve Climate Forecasts." *Nature* 548(7668): 379–80.

Jones, Michael D., and Mark K. McBeth. 2010. "A Narrative Policy Framework: Clear Enough to Be Wrong?" *Policy Studies Journal* 38(2): 329–53.

Kahan, Dan M. 2012. "Cultural Cognition as a Conception of the Cultural Theory of Risk." Eds. Roser, Sabine; Hillerbrand, Raffaela; Sandin, Per; Peterson, Martin. *Handbook of Risk Theory: Epistemology, Decision Theory, Ethics, and Social Implications of Risk*: 725–59.

Kaplan, Richard L. 2001. *Politics and the American Press: The Rise of Objectivity, 1865–1920*. Cambridge: Cambridge University Press.

Keller, Sarina. 2010. "Scientization: Putting Global Climate Change on the Scientific Agenda and the Role of the IPCC." *Poiesis und Praxis* 7(3): 197–209.

Keohane, Robert O., and David G. Victor. 2016. "Cooperation and Discord in Global Climate Policy." *Nature Climate Change* 6(6): 570–5.

Kintz, M. (2010). Die Berufsstruktur der Abgeordneten des 17. Deutschen Bundestages. *Zeitschrift Für Parlamentsfragen*, 41(3): 491–503.

Koppl, Roger. 2018. *Expert Failure*. Cambridge: Cambridge University Press.

Krajewski, Joanna M. T., and Brian Ekdale. 2017. "Constructing Cholera: CNN IReport, the Haitian Cholera Epidemic, and the Limits of Citizen Journalism." *Journalism Practice* 11(2–3): 229–46.

Krapohl, Sebastian. 2003. "Risk Regulation in the EU between Interests and Expertise: The Case of BSE." *Journal of European Public Policy* 10(2): 189–207.

Krick, Eva. 2021. "Citizen Experts in Participatory Governance : Democratic and Epistemic Assets of Service User Involvement, Local Knowledge and Citizen Science." *Current Sociology*.

Kumru, Seda, Pakize Yiğit, and Osman Hayran. 2021. "Demography, Inequalities and Global Health Security Index as Correlates of COVID-19 Morbidity and Mortality." *The International Journal of Health Planning and Management* 37(2):944–962.

Kupferschmidt, Kai. 2020. "Global Plan Seeks to Promote Vaccine Equity, Spread Risks." *Science* 369(6503): 489–90.
Lahn, Bård. 2020. "A History of the Global Carbon Budget." *Wiley Interdisciplinary Reviews: Climate Change* (December 2019): 1–9.
Latour, Bruno. 1983. "Give Me a Laboratory and I Will Raise the World." In *Science Observed*, eds. Karin Knorr-Cetina and Michael Mulkay. London: Sage, 141–70.
———. 1987. *Science in Action: How to Follow Scientists and Engineers Through Society*. Cambridge, MA: Harvard University Press.
———. 1993. *The Pasteurization of France*. Harvard University Press.
———. 2004. "Why Has Critique Run Out of Steam? From Matters of Fact to Matters of Concern." *Critical Inquiry* 30(2): 225–48.
———. 2005. *Reassembling the Social: An Introduction to Actor-Network-Theory*. Oxford: Oxford University Press.
———. 2007. "Turning Around Politics: A Note on Gerard de Vries' Paper." *Social Studies of Science* 37(5): 811–20.
———. 2013. *War and Peace in an Age of Ecological Conflicts*. Lecture prepared for the Peter Wall Institute Vancouver.
Latour, Bruno, and Steve Woolgar. 1986. *Laboratory Life: The Construction of Scientific Facts*. Princeton, NJ: Princeton University Press.
Lazer, David M. J. et al. 2018. "The Science of Fake News." *Science* 359(6380): 1094–96.
Le Sueur, Andrew. 2016. "Robot Government: Automated Decision-Making and Its Implications for Parliament." In *Parliament: Legislation and Accountability*, eds. Alexander Hart and Andrew Le Sueur. Oxford, UK: Hart Publishing, 183–202.
Leemans, Rik, and Pier Vellinga. 2017. "The Scientific Motivation of the Internationally Agreed 'Well below 2°C' Climate Protection Target: A Historical Perspective." *Current Opinion in Environmental Sustainability* 26–27: 134–42.
Lentsch, Justus, and Peter Weingart. 2009. *Scientific Advice to Policy Making: International Comparison*. Leverkusen, Opladen: Verlag Barbara Budrich.
———. 2011. "Quality Control in the Advisory Process: Towards an Institutional Design for Robust Science Advice." In *The Politics of Scientific Advice. Institutional Design for Quality Assurance*, eds. Justus Lentsch and Peter Weingart. Cambridge: Cambridge University Press, 353–74.
Levin, Andrew T. et al. 2020. "Assessing the Age Specificity of Infection Fatality Rates for COVID-19: Systematic Review, Meta-Analysis, and Public Policy Implications." *European Journal of Epidemiology* 35(12): 1123–38.
Lezaun, Javier, and Linda Soneryd. 2007. "Consulting Citizens: Technologies of Elicitation and the Mobility of Publics." *Public Understanding of Science* 16(3): 279–97.
Lidskog, Rolf, Ingemar Elander, and Adam Standring. 2020. "COVID-19, the Climate, and Transformative Change: Comparing the Social Anatomies of Crises and Their Regulatory Responses." *Sustainability (Switzerland)* 12(16): 1–21.
Lin, Zhiyuan Jerry, Jongbin Jung, Sharad Goel, and Jennifer Skeem. 2020. "The Limits of Human Predictions of Recidivism." *Science Advances* 6(7): 1–9.
Limoges, Camille. 1993. "Expert Knowledge and Decision-Making in Controversy Contexts." *Public Understanding of Science* 2(2): 417–26.
Lindblom, Charles. 1959. "The Science of 'Muddling Through.'" *Public Administration Review* 19(2): 79–88.
Lipset, Seymour Martin. 2001. "The Americanization of the European Left." *Journal of Democracy* 12(2): 74–87.

Liu, Xiaomo et al. 2016. "Reuters Tracer: A Large Scale System of Detecting & Verifying Real-Time News Events from Twitter." *International Conference on Information and Knowledge Management, Proceedings* 24–28 October, 207–16.

Lövbrand, Eva, Roger Pielke, and Silke Beck. 2010. "A Democracy Paradox in Studies of Science and Technology." *Science, Technology & Human Values* 36(4): 474–96.

Luhmann, Niklas. 2020. *Theory of Society*, Volume 1. Stanford, CA: Stanford University Press.

Lynch, Michael. 2016. "Social Constructivism in Science and Technology Studies." *Human Studies* 39(1): 101–12.

Madrigal, Alexis. 2014. "The Trick That Makes Google's Self-Driving Cars Work." *The Atlantic*. https://www.theatlantic.com/technology/archive/2014/05/all-the-world-a-track-the-trick-that-makes-googles-self-driving-cars-work/370871/ (January 22, 2020).

Mahony, Martin. 2013. "Boundary Spaces: Science, Politics and the Epistemic Geographies of Climate Change in Copenhagen, 2009." *Geoforum* 49: 29–39.

———. 2015. "Climate Change and the Geographies of Objectivity: The Case of the IPCC's Burning Embers Diagram." *Transactions of the Institute of British Geographers* 40(2): 153–67.

Majone, Giandomenico. 1989. *Evidence, Argument, and Persuasion in the Policy Process*. New Haven and London: Yale University Press.

Majone, G. 2005. Dilemmas of European Integration: The Ambiguities and Pitfalls of Integration by Stealth *Dilemmas of European Integration: The Ambiguities and Pitfalls of Integration by Stealth*. Oxford: Oxford University Press.

Mansbridge, Jane et al. 2012. "A Systemic Approach to Deliberative Democracy." In *Deliberative Systems*, eds. John Parkinson and Jane Mansbridge. Cambridge: Cambridge University Press, 1–26.

Marston, Cicely, Alicia Renedo, and Sam Miles. 2020. "Community Participation Is Crucial in a Pandemic." *The Lancet 4 May*.

Marx, Karl. 1975. "A Contribution to the Critique of Hegel's Philosophy of Right 1844." In *Collected Works*, Vol. 3, London: Lawrence & Wishart.

Masson-Delmotte, V. et al. 2021. *Climate Change 2021: The Physical Science Basis. Summary for Policymakers (SPM)*. Cambridge: Cambridge University Press.

Mayer-Schönberger, Viktor, and Kenneth Cukier. 2013. *Big Data: A Revolution That Will Transform How We Live, Work, and Think*. Boston, MA: Houghton Mifflin Harcourt.

Mazzucato, Mariana. 2011. *The Entrepreneurial State: Debunking Public vs. Private Sector Myths*. London: Anthem Press.

McCright, A.M., and R.E. Dunlap. 2003. "Defeating Kyoto: The Conservative Movement's Impact on US Climate Change Policy." *Social Problems* 50(3): 348–73.

McLaren, Duncan, and Nils Markusson. 2020. "The Co-Evolution of Technological Promises, Modelling, Policies and Climate Change Targets." *Nature Climate Change* 10(5): 392–97.

Meehl, Paul E. 1954. *Clinical versus Statistical Prediction: A Theoretical Analysis and a Review of the Evidence*. Minneapolis, MN: University of Minnesota Press.

Milkoreit, Manjana, and Kate Haapala. 2017. "Designing the Global Stocktake: A Global Governance Innovation." Working Paper, Arlington, VA: Centre for Climate and Energy Solutions.

Miller, Clark A. 2004. "Climate Science and the Making of a Global Political Order." In *States of Knowledge: The Co-Production of Science and the Social Order*, ed. Sheila Jasanoff. London and New York: Routledge, 46–66.

Minx, Jan C. et al. 2017. "Learning about Climate Change Solutions in the IPCC and Beyond." *Environmental Science and Policy* 77(October 2016): 252–9.

Mouffe, Chantal. 2005. *On the Political*. Abingdon: Routledge.

Murphy, Yvonne. 2016. "The Marriage Equality Referendum 2015." *Irish Political Studies* 31(2): 315–30.

Narayanan, Arvind. 2019. "How to Recognize AI Snake Oil." https://arxiv.org/abs/1906.09208 (November 21, 2019).

Nowotny, Helga. 2021. *In AI We Trust: Power, Illusion and Control of Predictive Algorithms*. Cambridge: Polity.

O'Gorman, Paul. 2019. "3Q: Machine Learning and Climate Modeling." *MIT News*. https://news.mit.edu/2019/mit-3q-paul-o-gorman-machine-learning-for-climate-modeling-0213.

ONS. 2021. "Prevalence of Ongoing Symptoms Following Coronavirus (COVID-19) Infection in the UK—Office for National Statistics." https://www.ons.gov.uk/peoplepopulationandcommunity/healthandsocialcare/conditionsanddiseases/bulletins/prevalenceofongoingsymptomsfollowingcoronaviruscovid19infectionintheuk/1april2021 (January 19, 2022).

Oreskes, Naomi. 2019. *Why Trust Science?* Princeton, NJ: Princeton University Press.

Oreskes, Naomi, and Erik M. Conway. 2012. *Merchants of Doubt: How a Handful of Scientists Obscured the Truth on Issues from Tobacco Smoke to Global Warming*. London: Bloomsbury Paperbacks.

Oreskes, Naomi, Michael Oppenheimer, and Dale Jamieson. 2019. "Scientists Have Been Underestimating the Pace of Climate Change." *Scientific American Blog Network*. https://blogs.scientificamerican.com/observations/scientists-have-been-underestimating-the-pace-of-climate-change/ (December 8, 2019).

Oswald, Marion. 2018. "Algorithm-Assisted Decision-Making in the Public Sector: Framing the Issues Using Administrative Law Rules Governing Discretionary Power." *Philosophical Transactions of the Royal Society A: Mathematical, Physical and Engineering Sciences* 376:1–20.

Oxfam International. 2021. "Vaccine Monopolies Make Cost of Vaccinating the World against COVID at Least 5 Times More Expensive than It Could Be." https://www.oxfam.org/en/press-releases/vaccine-monopolies-make-cost-vaccinating-world-against-covid-least-5-times-more (September 6, 2021).

Paglia, Eric, and Charles Parker. 2021. "The Intergovernmental Panel on Climate Change: Guardian of Climate Science." In *Guardians of Public Value: How Public Organisations Become and Remain Institutions*, eds. Arjen Boin, Lauren A Fahy, and Paul 't Hart. Cham: Springer International Publishing, 295–321.

Painter, James. 2015. "The Effectiveness of the IPCC Communication: A Survey of (Mainly) UK-Based Users." *Joint Programming Initiative (JPI) AR5 in Europe*: 1–98.

Papadopoulos, Dimitris. 2018. *Experimental Practice: Technoscience, Alterontologies, and More-than-Social Movements*. Durham, NC: Duke University Press.

Parthasarathy, Shobita. 2010. "Breaking the Expertise Barrier: Understanding Activist Strategies in Science and Technology Policy Domains." *Science and Public Policy* 37(5): 355–67.

Pasquale, Frank. 2015. *The Black Box Society: The Secret Algorithms That Control Money and Information*. Cambridge, MA: Harvard University Press.

Pearce, Warren et al. 2017. "Beyond Counting Climate Consensus." *Environmental Communication* 11(6): 723–30.

———. 2020. "Trouble in the Trough: How Uncertainties Were Downplayed in the UK's Science Advice on Covid-19." *Humanities and Social Sciences Communications*.

Pearse, Harry. 2020. "Deliberation, Citizen Science and Covid-19." *The Political Quarterly* 91(3): 571–7.

Peiser, Jaclyn. 2019. "The Rise of the Robot Reporter." *New York Times*. 5 February.

Pellizzoni, Luigi. 2001. "The Myth of the Best Argument: Power, Deliberation and Reason." *British Journal of Sociology* 52(1): 59–86.

Peng, Wei et al. 2021. "Climate Policy Models Need to Get Real about People—Here's How." *Nature* 594:174–76.

Peters, Glen P. 2018. "Beyond Carbon Budgets." *Nature Geoscience* 11(6): 378–80.

Peters, Glen P., and Zeke Hausfather. 2020. "Emissions—the 'Business as Usual' Story Is Misleading." *Nature* 577: 618–20.

Peters, B. Guy. 2017. "What Is so Wicked about Wicked Problems? A Conceptual Analysis and a Research Program." *Policy and Society* 36(3): 385–96.

Pielke Jnr., Roger. 2007. *The Honest Broker*. Cambridge: Cambridge University Press.

———. 2010. *The Climate Fix*. New York: Basic Books.

Pielke, Roger Jr. 2010. "Major Change Is Needed If the IPCC Hopes to Survive." *Yale E360*. https://e360.yale.edu/features/major_change_is_needed_if_the_ipcc_hopes_to_survive (December 8, 2019).

Pielke, Roger, Gwyn Prins, Steve Rayner, and Daniel Sarewitz. 2007. "Climate Change 2007: Lifting the Taboo on Adaptation." *Nature* 445(7128): 597–8.

Pindyck, Robert S. 2017. "The Use and Misuse of Models for Climate Policy." *Review of Environmental Economics and Policy* 11(1): 100–14.

Pirro, A. L. P., P. Taggart, and S. van Kessel. 2018. "The Populist Politics of Euroscepticism in Times of Crisis: Comparative Conclusions." *Politics* 38(3): 378–90.

Prainsack, Barbara. 2019. "Data Donation: How to Resist the iLeviathan." In *The Ethics of Medical Data Donation. Philosophical Studies Series*, Vol. 137, eds. J. Krutzinna and L. Floridi. Cham: Springer, 9–22.

Prior, Lindsay. 2003. "Belief, Knowledge and Expertise: The Emergence of the Lay Expert in Medical Sociology." *Sociology of Health & Illness* 25(3): 41–57.

Przeworski, Adam. 1999. "Minimalist Conception of Democracy: A Defense." In *Democracy's Value*, eds. Ian Shapiro and Casiano Hacker-Cordon. Cambridge: Cambridge University Press, 23–55.

Rabeharisoa, Vololona, and Michel Callon. 2004. "Patients and Scientists in French Muscular Dystrophy Research." In *States of Knowledge: The Co-Production of Science and the Social Order*, ed. Sheila Jasanoff. London: Routledge, 142–60.

Raghavan, Manish, Solon Barocas, Jon Kleinberg, and Karen Levy. 2019. "Mitigating Bias in Algorithmic Employment Screening: Evaluating Claims and Practices." *SSRN Electronic Journal*: 1–24.

Ramírez-i-Ollé, Meritxell. 2015. "Rhetorical Strategies for Scientific Authority: A Boundary-Work Analysis of 'Climategate.'" *Science as Culture* 5431(March): 1–28.

Rammert, Werner. 2008. "Where the Action Is Distributed Agency Between Humans, Machines, and Programs." In *Paradoxes of Interactivity*, eds. Uwe Seifert, Jin Huyn Kim, and Anthony Moore. New Brunswick, NJ: Transaction Publishers, 63–91.

Randalls, Samuel. 2010. "History of the 2° C Climate Target." *Wiley Interdisciplinary Reviews: Climate Change* 1(4): 598–605.

Ravetz, Jerome. 1999. "What Is Post-Normal Science." *Futures* 31: 647–53.

———. 2011. "'Climategate' and the Maturing of Post-Normal Science." *Futures* 43(2): 149–57.

Rayner, Steve. 2006. "Wicked Problems: Clumsy Solutions—Diagnoses and Prescriptions for Environmental Ills." *Jack Beale Memorial Lecture on Global Environment* (July): 1–12.

Reicher, Stephen, and Clifford Stott. 2020. "On Order and Disorder during the COVID-19 Pandemic." *British Journal of Social Psychology* 59(3): 694–702.

Renn, Ortwin. 1999. "Policy Analysis A Model for an Analytic—Deliberative Process in Risk Management." 33(18):3049–55.

Richta, Radovan. 1969. *Civilization at the Crossroads. Social and Human Implications of the Scientific and Technological Revolution*. White Plains, NY: International Arts and Sciences Press.

Riker, William H. 1986. *The Art of Political Manipulation*. New Haven, CT: Yale University Press.

Rittberger, B. 2005. *Building Europe's Parliament: Democratic Representation Beyond the Nation State*. Oxford: Oxford University Press.

Rittel, Horst, and Melvin M Webber. 1973. "Dilemmas in a General Theory of Planning." *Policy Sciences* 4: 155–69.

Rogelj, Joeri et al. 2018. "Scenarios towards Limiting Global Mean Temperature Increase below 1.5°C." *Nature Climate Change* 8(4): 325–32.

Rogelj, Joeri, Malte Meinshausen, and Reto Knutti. 2012. "Global Warming under Old and New Scenarios Using IPCC Climate Sensitivity Range Estimates." *Nature Climate Change* 2(4): 248–53.

Ross, Casey, and Ike Swetlitz. 2018. "IBM's Watson Recommended 'Unsafe and Incorrect' Cancer Treatments." *STAT*. https://www.statnews.com/2018/07/25/ibm-watson-recommended-unsafe-incorrect-treatments/ (November 26, 2021).

Roszak, Theodore. 1968. *The Making of a Counter Culture: Reflections on the Technocratic Society and Its Youthful Opposition*. Berkeley, CA: University of California Press.

Royal Historical Society. 2016. "Addressing Authority: An Online Symposium on Petitions and Supplications in Early Modern Society." https://royalhistsoc.org/calendar/addressing-authority-online-symposium-petitions-supplications-early-modern-society/

Rubin, Olivier, Nicole A. Errett, Ross Upshur, and Erik Baekkeskov. 2021. "The Challenges Facing Evidence-Based Decision Making in the Initial Response to COVID-19." *Scandinavian Journal of Public Health* 49(7):790–796.

Runciman, David. 2019. *How Democracy Ends*. London: Profile Books.

Saloranta, Tuomo. 2001. "Post-Normal Science and the Global Climate Change Issue." *Climatic Change* 50: 396–404.

Salter, Liora, Edwin Levy, and William Leiss. 1988. *Mandated Science: Science and Scientists in the Making of Standards*. Dordrecht: Springer.

Sarewitz, Daniel. 2000. "Science and Environmental Policy. An Excess of Objectivity." In *Earth Matters: The Earth Sciences, Philosophy and the Claims of Community*, ed. R. Frodeman. Upper Saddle River, NJ: Prentice Hall, 79–98.

Sarewitz, Daniel, and Richard Nelson. 2008. "Three Rules for Technological Fixes." *Nature* 456(7224): 871–2.

Sartor, G., and L. Karl Branting. 1998. "Introduction: Judicial Applications of Artificial Intelligence." *Artificial Intelligence and Law* 6(2–4): 105–10.

Schattschneider, Elmer Eric. 1960. *The Semi-Sovereign People*. New York: Holt, Rinehart and Winston.

Schiffer, Steven Jay. 2018. "'Glocalized' Utopia, Community-Building, and the Limits of Imagination." *Utopian Studies* 29(1): 67–87.

Schneider, Stephen. 1991. "Three Reports of the Intergovernmental Panel on Climate Change." *Environment* 33(1): 25–30.

Schudson, Michael. 1978. *Discovering the News: A Social History of American Newspapers*. New York: Basic Books.

Schumacher, E. F. 1971. *Small Is Beautiful. A Study of Economics as If People Mattered*. London: Blond and Briggs.

Shackley, Simon, Peter Young, Stuart Parkinson, and Brian Wynne. 1998. "Uncertainty, Complexity and Concepts of Good Science in Climate Change Modelling: Are GCMs the Best Tools?" *Climatic Change* 38(2): 159–205.

Shapin, Steven. 2007. "Expertise, Common Sense, and the Atkins Diet." *Public Science in Liberal Democracy* ed. J. Porter and P. W. B. Phillips, 174–193. Toronto: University of Toronto Press, 174–93.

Shiller, Robert J. 2017. "Narrative Economics." *American Economic Review* 107(4): 967–1004.

Siebenhüner, Bernd. 2003. "The Changing Role of Nation States in International Environmental Assessments—the Case of the IPCC." *Global Environmental Change* 13(2): 113–23.

Simmons, Michael. 2022. "Sage Scenarios vs Actual: An Update." *The Spectator*. 23 January.

Simon, Herbert A. 1973. "The Structure of Ill Structured Problems." *Artificial Intelligence* 4(3–4): 181–201.

Skodvin, Tora. 2000. *Structure and Agent in the Scientific Diplomacy of Climate Change: An Empirical Case Study of Science–Policy Interaction in the Intergovernmental Panel on Climate Change*. Dordrecht: Springer.

Skolnikoff, Eugene B. 1997. *Same Science, Differing Policies; The Saga of Global Climate Change*. The MIT Joint Program on the Science and Policy of Global Change Joint Program Report Series, 19 pages.

Smith, Philip, and Nicolas C. Howe. 2015. *Climate Change as Social Drama: Global Warming in the Public Sphere*. Cambridge: Cambridge University Press.

Sniezek, Janet A., Gunnar E. Schrah, and Reeshad S. Dalal. 2004. "Improving Judgement with Prepaid Expert Advice." *Journal of Behavioral Decision Making* 17(3): 173–90.

Snowden, Edward. 2019. *Permanent Record*. London: Metropolitan Books.

Specia, Megan. 2022. "The Omicron Shift in Europe: Pandemic or Endemic?" *New York Times*, 14 January.

Speth, Rudolf. 2016. "Lobbying Und Expertise. Wie Sich Politik Expertise Verschafft." In *Auf Dem Markt Der Experten. Zwischen Überforderung Und Vielfalt*, ed. Pfizenmaier, Ruben. Frankfurt am Main: Edition Büchergilde, 72–88.

Spiegelhalter, David, and Anthony Masters. 2021. "Will We Reach Herd Immunity for the New Coronavirus?" *The Guardian*, 14 August.

Splichal, S., and P. Dahlgren. 2016. "Journalism between De-Professionalisation and Democratisation." *European Journal of Communication* 31(1): 5–18.

Sridhar, Shrihari, and Srinivasaraghavan Sriram. 2015. "Is Online Newspaper Advertising Cannibalizing Print Advertising?" *Quantitative Marketing and Economics* 13(4): 283–318.

Stehr, Nico. 2015. "Climate Policy: Democracy Is Not an Inconvenience." *Nature* 525(7570): 449–50.

Stehr, Nico, and Reiner Grundmann. 2011. *Experts: The Knowledge and Power of Expertise*. London: Routledge.

Stephens, Bret. 2021. "Covid Misinformation Comes From the Top, Too." *New York Times* 27. July.

Stevens, Alex. 2020. "Governments Cannot Just 'Follow the Science' on COVID-19." *Nature Human Behaviour* 4(6): 560.

Stilgoe, Jack. 2020. *Who's Driving Innovation?*London: Palgrave.

Stirling, Andrew. 2008. "'Opening Up' and 'Closing Down': Power, Participation, and Pluralism in the Social Appraisal of Technology." *Science Technology and Human Values* 33(2): 262–94.

Stone, Deborah A. 2002. *Policy Paradox: The Art of Political Decision Making*. Revised 3edn. New York: Norton.

Stone, Diane, and Andrew Denham. 2004. *Think Tank Traditions: Policy Analysis Across Nations*. Manchester: Manchester University Press.

Strickland, Eliza. 2019. "How IBM Watson Overpromised and Underdelivered on AI Health Care—IEEE Spectrum." *IEEE Spectrum*. https://spectrum.ieee.org/how-ibm-watson-overpromised-and-underdelivered-on-ai-health-care (November 26, 2021).

Suh, Jong Hwan, Chung Hoon Park, and Si Hyun Jeon. 2010. "Applying Text and Data Mining Techniques to Forecasting the Trend of Petitions Filed to E-People." *Expert Systems with Applications* 37(10): 7255–68.

Sundqvist, Göran et al. 2018. "One World or Two? Science–Policy Interactions in the Climate Field." *Critical Policy Studies* 12(4): 448–68.

Susskind, Richard, and Daniel Susskind. 2016. *The Future of the Professions. How Technology Will Transform the Work of Human Experts*. Oxford: Oxford University Press.

Szöcsik, E, and A Polyakova. 2018. "Euroscepticism and the Electoral Success of the Far Right: The Role of the Strategic Interaction between Center and Far Right." *European Political Science*: 18: 400–420.

Taylor, Russell H., and Dan Stoianovici. 2003. "Medical Robotics in Surgery." *IEEE Transactions on Robotics And Automation* 19(5): 765–81.

Tetlock, Philip E. 2005. *Expert Political Judgment : How Good Is It? How Can We Know?* Princeton, NJ: Princeton University Press.

———. 2010. "Second Thoughts about Expert Political Judgment: Reply to the Symposium." *Critical Review* 22(4): 467–88.

Tetlock, Philip, and Dan Gardner. 2015. *Superforecasting : The Art and Science of Prediction*. New York: Crown.

Teubner, Gunther. 2006. "Rights of Non-Humans? Electronic Agents and Animals as New Actors in Politics and Law." *Journal of Law and Society* 33(4): 497–521.

The News 24. 2021. "Pfizer's Secret Contract Leaked—Customers Are at the Mercy of the Pharmaceutical Giant." 4 August. https://then24.com/2021/08/04/pfizers-secret-contract-leaked-customers-are-at-the-mercy-of-the-pharmaceutical-giant/ (September 6, 2021).

Thompson, Dennis F. 2008. "Deliberative Democratic Theory and Empirical Political Science." *Annual Review of Political Science* 11(1): 497–520.

Thunert, Martin. 2004. "Think Tanks in Germany." *Society* 41: 66–69.

Tol, Richard S. J. 2010. ESRI Working Paper: *Regulating Knowledge Monopolies: The Case of the IPCC*. Dublin.

Trentmann, Frank, Anna Barbara Sum, Manuel Rivera, and Gesellschaft für Ökologische Kommunikation mbH. 2019. *Work in Progress Economy and Environment in the Hands of Experts*. Munich: Oekonom.

Turchetti, Simone, and Peder Roberts. 2015. *The Surveillance Imperative: Geosciences during the Cold War and Beyond*. New York: Palgrave Macmillan.

Turner, Stephen. 2001. "What Is the Problem with Experts?" *Social Studies of Science* 31(1): 123–49.

———. 2003. *Liberal Democracy 3.0: Civil Society in an Age of Experts*. London: SAGE Publications.

Turner, Stephen P. 2010. "Normal Accidents of Expertise." *Minerva* 48(3): 239–58.

———. 2019. "Circles or Regresses? The Problem of Genuine Expertise, Stephen." *Social Epistemology Review and Reply Collective*. https://social-epistemology.com/2019/04/11/circles-or-regresses-the-problem-of-genuine-expertise-stephen-turner/ (December 1, 2021).

Turnpenny, John, Mavis Jones, and Irene Lorenzoni. 2011. "Where Now for Post-Normal Science?: A Critical Review of Its Development, Definitions, and Uses." *Science, Technology, & Human Values* 36(3): 287–306.

UNFCCC. 1992. *United Nations Framework Convention on Climate Change*. https://unfccc.int/resource/docs/convkp/conveng.pdf

United Nations. 1988. UNGA Resolution on the Protection of Global Climate for Present and Future Generations of Mankind, UN Doc. A/RES/43/53, 6 December 1988.

van der Linden, Sander, Jon Roozenbeek, and Josh Compton. 2020. "Inoculating Against Fake News About COVID-19." *Frontiers in Psychology* 11: 2928.

Vanderstraeten, Raf. 2002. "Parsons, Luhmann and the Theorem of Double Contingency." *Journal of Classical Sociology* 2(1): 77–92.

van Vuuren, Detlef P. et al. 2011. "The Representative Concentration Pathways: An Overview." *Climatic Change* 109(1–2): 5–31.

Verweij, M. et al. 2006. "Clumsy Solutions for a Complex World: The Case of Climate Change." *Public Administration* 84(4): 817.

Vibert, Frank. 2007. *The Rise of the Unelected: Democracy and the New Separation of Powers*. Cambridge: Cambridge University Press.

Vogel, Gretchen, and Kai Kupferschmidt. 2021. "'It's a Very Special Picture.' Why Vaccine Safety Experts Put the Brakes on AstraZeneca's COVID-19 Vaccine." *Science* (17 March).

Wagner, Gerhard. 2019. "Robot, Inc.: Personhood for Autonomous Systems?" *Fordham Law Review* 88(2): 591–612.

Walsh, Lynda. 2013. *Scientists as Prophets*. New York: Oxford University Press.

Waltz, Emily. 2020. "What AI Can–and Can't–Do in the Race for a Coronavirus Vaccine." *IEEE Spectrum*. https://spectrum.ieee.org/what-ai-can-and-cant-do-in-the-race-for-a-coronavirus-vaccine (November 28, 2021).

Warren, Mark E. 1996. "Deliberative Democracy and Authority." *The American Political Science Review* 90(1): 46–60.

Waters, Richard. 2020. "Big Tech Searches for a Way Back into Healthcare." *Financial Times*, 17 May.

Watson, Hayley. 2011. "Preconditions for Citizen Journalism: A Sociological Assessment." *Sociological Review Online* 16(3):82–93.

Watson, Jamie Carlin. 2019. "What Experts Could Not Be." *Social Epistemology* 33(1): 74–87.

Weaver, C. P. et al. 2014. "From Global Change Science to Action with Social Sciences." *Nature Climate Change* 4(8): 656–9.

Weber, Max. 1978. *Economy and Society: An Outline of Interpretive Sociology*. Berkeley, CA: University of California Press.

———. 2012. "On Some Categories Interpretive Sociology." In *Max Weber: Collected Methodological Writings*, eds. Hans Henrik Bruun and Sam Whimster. London: Routledge, 273–301.

Weinberg, Alvin M. 1972. "Science and Trans-Science." *Minerva* 10(2): 209–22.

White, Hayden. 1980. "The Value of Narrativity in the Representation of Reality." *Critical Inquiry* 7(1): 5–27.

WHO. 2020. *Strengthening and Adjusting Public Health Measures throughout the COVID-19 Transition Phases Policy Considerations for the WHO European Region*. https://www.euro.who.int/__data/assets/pdf_file/0019/440038/StrengthAdjustingMeasuresCOVID19-infograph.pdf

Wille, Joachim. 2018. "Die Haut Und Die Freiheit Retten." *Frankfurter Rundschau*: 3. September.

World Health Organization. 2005. "WHO Constitution." https://apps.who.int/gb/bd/ (October 13, 2020).

Wynne, Brian. 1989. "Sheepfarming after Chernobyl: A Case Study in Communicating Scientific Information." *Environment: Science and Policy for Sustainable Development* 31(March 2015): 10–39.

———. 1996. "May the Sheep Safely Graze? A Reflexive View of the Expert–Lay Knowledge Divide." *In Risk, Environment and Modernity: Towards a New Ecology*, eds. Scott Lash, Branislaw Szerszynski, and Brian Wynne. London: Sage, 44–83.

Yearley, Steven. 2000. "Making Systematic Sense of Public Discontents with Expert Knowledge: Two Analytical Approaches and a Case Study." *Public Understanding of Science* 9: 105–22.

———. 2009. "Sociology and Climate Change after Kyoto: What Roles for Social Science in Understanding Climate Change?" *Current Sociology* 57(3): 389–405.

Young, Iris Marion. 2001. "Activist Challenges to Deliberative Democracy:" *Political Theory* 29(5): 670–90.

Zorita, Eduardo. 2019. "The Climate of the Past Millennium and Online Public Engagement in a Scientific Debate." *Wiley Interdisciplinary Reviews: Climate Change* 10(5): e590.

Zuboff, Shoshana. 2019. *The Age of Surveillance Capitalism: The Fight for the Future at the New Frontier of Power*. London: Profile Books.

Index

For Product Safety Concerns and Information please contact our EU
representative GPSR@taylorandfrancis.com
Taylor & Francis Verlag GmbH, Kaufingerstraße 24, 80331 München, Germany

www.ingramcontent.com/pod-product-compliance
Ingram Content Group UK Ltd.
Pitfield, Milton Keynes, MK11 3LW, UK
UKHW022353190926
475702UK00015B/2272

* 9 7 8 1 0 3 2 3 3 5 6 4 3 *